TEN DRUGS

TEN DRUGS

HOW PLANTS, POWDERS, AND PILLS HAVE
SHAPED THE HISTORY OF MEDICINE

by Thomas Hager

ABRAMS PRESS, NEW YORK

Library of Congress Control Number: 2018936303

ISBN: 978-1-4197-3440-3
eISBN: 978-1-68335-531-1

Printed and bound in the United States

10 9 8 7 6 5 4 3 2 1

Abrams books are available at special discounts when purchased in quantity
for premiums and promotions as well as fundraising or educational use.
Special editions can also be created to specification. For details, contact
specialsales@abramsbooks.com or the address below.

Abrams Press® is a registered trademark of Harry N. Abrams, Inc.

ABRAMS The Art of Books
195 Broadway, New York, NY 10007
abramsbooks.com

For Jackson, Zane, and Elizabeth

CONTENTS

Introduction   *50,000 Pills*   1

CHAPTER 1   *The Joy Plant*   11
CHAPTER 2   *Lady Mary's Monster*   49
CHAPTER 3   *The Mickey Finn*   75
CHAPTER 4   *How to Soothe Your Cough with Heroin*   85
CHAPTER 5   *Magic Bullets*   99
CHAPTER 6   *The Least Explored Territory on the Planet*   123

INTERLUDE   *THE GOLDEN AGE*   159

CHAPTER 7   *Sex, Drugs, and More Drugs*   163
CHAPTER 8   *The Enchanted Ring*   187
CHAPTER 9   *Statins: A Personal Story*   211
CHAPTER 10   *A Perfection of Blood*   241

Epilogue   *The Future of Drugs*   259

Source Notes   271
Bibliography   277
Index   287

50,000 PILLS

ON A BUSINESS TRIP years ago, I had an extra day in London. So like many tourists, I headed to the British Museum. And there I ran across something extraordinary.

In a large, light-filled gallery on the ground floor was a table covered with thousands of pills. It was an exhibit conceived by an artist and a doctor who had come up with a way to display all the 14,000 doses of prescription drugs an average Briton took in a lifetime. These pills, woven into lengths of fabric and accompanied by bits of explanatory text, covered a gallery table that stretched forty-six feet. I couldn't believe what I was seeing. Did people really take this many pills?

The answer is: No. They take more. The display was geared for Britain. And when it comes to taking drugs, the British don't come close to Americans. More than half of all Americans take at least one prescription drug on a regular basis, and most of those who fall into that group take more than one (somewhere between four and twelve prescriptions per person per year, depending on which study you look at). One expert estimates that Americans take an average of ten pills per person per day. Add in nonprescription drugs—over-the-counter vitamins, cold and flu remedies, aspirin, and other supplements—and run the numbers: Let's say a low-ball estimate of two pills per day per American over an average seventy-eight-plus years of life. The total comes to somewhere more than 50,000 pills, on average, in the average American's lifetime. And there's a good chance it's a lot more. America consumes more pharmaceuticals than any other nation on earth, and we spend a lot more to get them: more than $34 billion each year on over-the-counter drugs, and $270 billion on prescription drugs. That's way beyond what any other nation spends, because our drug prices are a lot higher than any other nation's. Americans constitute less than 5 percent of the world population but spend more than 50 percent of the money that flows into the world's drug companies.

And that's not even counting illegal drugs.

No nation in human history has taken as many drugs or spent as much money to get them as the United States does today. And the drugs have had profound effects. They have added decades to our average life spans, playing a central role in the graying of America. Drugs have changed the social and professional options of women. Drugs have altered the ways we view our minds, changed our attitudes toward the law, shifted international relations, and triggered wars.

By these measures, perhaps we should rename our species *Homo pharmacum*, the species that makes and takes drugs. We are the People of the Pill.

This book will introduce you to how we got here, with a focus on medical (that is, legal, non-recreational, mostly prescription) drugs. It is written as a series of brief, vivid sketches, sort of mini-biographies of ten drugs that changed medical history, linked by common themes, with each story leading into the next.

One of those common themes is the evolution of drugs. The

word *drug* itself comes from old French and Dutch terms for the barrels once used to keep herbs dry. Pharmacists 150 years ago were in many ways like today's herbalists, extracting and compounding their medicines for the most part from jars of dried plants. That gave doctors in the 1800s a couple-dozen somewhat effective natural medicines to help their patients (along with hundreds of useless, often alcohol-rich elixirs, poultices, and pills made and ballyhooed by local pharmacists). Today we have ten-thousand-plus, ever-more-targeted, increasingly powerful high-tech medicines that can treat and often cure conditions that have confounded healers for thousands of years.

Wrapped in this evolution and guiding its trajectory is humanity's search for magic bullets, medicines that can unerringly seek out and destroy diseases in our bodies without doing any harm to our health along the way. The goal has always been to find medicines that are all-powerful, but without any risk. That is likely an impossible goal. We haven't yet found a perfect magic bullet. But we keep inching closer.

Another thread that runs through these chapters tells a bit about the growth of the industry that makes drugs—the trillion-dollar behemoth that critics have dubbed "Big Pharma"—and changes in the ways we regulate that industry. For instance, in the 1880s you could get just about any drug you wanted without a prescription, over the counter, including mixtures laced with opium, cocaine, and cannabis. Now you need a prescription for almost any powerful medicinal drug, and even with a prescription you can't buy narcotics like heroin (well, at least not in the United States). Drugmakers before 1938 could put just about anything they wanted on the market as long as it didn't kill you, and they didn't try to fool you with false advertising. Today, prescription drugs have to be proven both safe and effective before they can be sold. These laws governing our drugs evolved, in sometimes surprising ways, along with the drugs themselves.

Our attitudes have changed, too. In the 1880s, most people considered the right to self-medicate as something close to inalienable. It didn't matter if a drug was good for you or not, deciding whether to take it was your choice, not your doctor's. If you wanted to buy one of the many patent medicine horrors available from local drugstores, anything from radioactive water for cancer to opium-spiked syrups for insomnia, well, it was your body. Nobody had the right to tell you otherwise.

Today that's been turned on its head. Now physicians hold the keys (in the form of their prescription pads) to getting most drugs. Today, when it comes to taking our medicine, we pretty much do as we're told.

Drugs changed the practice of medicine, too. In the 1880s, doctors were family counselors good at diagnosing ailments and providing solace and advice to relatives, but almost powerless to alter the course of killer diseases. Today, physicians are able to work miracles of lifesaving that their brethren a century ago could only dream of. They are also all too often overscheduled, data-stuffed technocrats more comfortable reading lab reports than holding a patient's hand.

During the past sixty years, the average life expectancy for Americans has lengthened by two months each year—mostly because of drugs. Vaccines have allowed us to completely conquer age-old enemies like smallpox (and we're closing in on polio). Prescription drugs, along with public health efforts, have made our lives much longer and, in general, healthier.

Not that there aren't also great risks. Drug overdoses, from both legal and illegal sources, are killing around 64,000 people each year, an annual death toll that exceeds all U.S. military deaths in all the years of the Vietnam War.

Here's what drugs have done for us: In the bad old days, say two hundred years ago, men lived twice as long, on average, as women (mostly because of the dangers of childbearing and -birth). And everybody in general lived about half as long as they do today. A lot of that was tied to death early in life. If babies made it through the risks and traumas of childbirth, survived the epidemic diseases of childhood—smallpox, measles, whooping cough, diphtheria, and more—and made it to adulthood, they could be considered lucky. Because then they could die of consumption, quinsy, cholera, erysipelas, gangrene, dropsy, syphilis, scarlet fever, or any of a few dozen other diseases that we don't hear much about anymore. Today we die from heart disease and cancer, diseases of the middle-aged and elderly. People in the old days didn't worry too much about heart disease or cancer because few people in the old days lived long enough to get them. Thanks to drugs, a group of scientists recently wrote, "People have different diseases, doctors hold different ideas about those diseases, and diseases carry different meanings in society."

As you'll see in this book, vaccines and antibiotics moved us from being helpless victims of epidemics to being able to fight them off. Combined with more effective public health measures—cleaner drinking water, better sewage systems, better hospitals—drugs moved us from fearing the diseases of childhood to suffering the diseases of the old. That's a tribute to medicine in general, and to drugs in particular.

These are technological tools capable of changing our culture. But when you think about them, drugs are even stranger than that. Today's pharmaceuticals are high-tech, developed in cutting-edge laboratories after investments of tens of millions of dollars, but a kind of high-tech so intimate, so personal, that they have to become part of you to do their work. You have to snort them, drink them, ingest them, inject them, rub them into your skin, make them part of your body. They dissolve inside you and race through your blood from muscle to heart, liver to brain. Only then, when they are absorbed, when they have melted into you and melded with you, does their power unfold. Then they can attach and trigger, soothe and calm, destroy and protect, alter your consciousness, restore your health. They can jack you up or chill you out. They can addict you, and they can save your life.

What gives them this power? Are they animal, vegetable, or mineral? All of the above. Are they good for you? Often. Are they dangerous? Always. Can they perform miracles? They can. Can they enslave us? Some do.

SO, EVER-MORE POWERFUL drugs, ever-more powerful physicians, ever-more diseases conquered. Seen this way, the story of drugs looks like a triumphant march of progress. But don't be fooled: Much of the history of drugs, as you'll see, is rooted in error, accidents, and lucky breaks.

Writing this book has, however, also convinced me that good old-fashioned progress plays a central role, too, if you define progress as the logical, rational application of a growing number of tested facts. Each new drug tells us new things about the body, and each new understanding of the body allows us to make better drugs. When the system is working well, each new scientific finding is criticized, tested and retested, amended if necessary, and then becomes part of a global library of facts available to other scientists. It builds. This synergy between drug-making and basic science, this dance between lab and pill and body, described in tens of thousands of scientific publications over the past three centuries, is now speeding up in tempo and growing in intensity. It is truly progressive. If we can hold our world together, we are on the brink of greater things.

I'LL TELL YOU what this book is not.

It is not a scholarly history of the pharmaceutical industry. It contains no footnotes. It ignores—out of necessity, for brevity—many world-shaking drug developments. You won't find every important drug here. But you'll find many of the drugs that have shaped both medical history and today's world. I hope you'll come away with a better understanding of this fascinating part of society.

It is not a book that will teach drug scientists anything very new, because it was not written for drug scientists. Rather, this book is for people who know just a little about drugs and want to learn more. It is aimed at the general reader, not the specialist—although I hope specialists, too, might come away with some interesting new stories to tell.

It is not a book that will make drug manufacturers happy. Or pro-pharma lobbyists. Or anti-pharma activists. It is neither a screed about the evils of the drug industry nor is it a song of praise for the wonders of science. I have no ax to grind, no agenda to promote.

My hope simply is to entertain you and to introduce you to a new world—the world of drug discovery—in a way that explains not only a lot about the history of medicine, but also something about our lives today, from our relationship with our doctors to the advertisements we see on TV, from the epidemic of opioid abuse to the possibilities of personalized medicine. Drug companies make incredible profits, and yet many of us can't afford the drugs we need. This book will get you thinking about why.

IF THERE'S ONE overarching lesson I hope you come away with, it is this: No drug is good. No drug is bad. Every drug is both.

Another way of saying that is that every effective drug, without exception, also comes with potentially dangerous side effects. This can be easy to forget in the first flush of enthusiasm when a new drug hits the market. Pushed by huge ad campaigns and often buttressed by glowing media reports, newly released blockbuster drugs enter what's called the Seige cycle (named after Max Seige, a

German researcher who first described it early in the last century). It happens time and time again: An astounding new drug is released to intense enthusiasm and wide adoption (that's stage 1 of the Seige cycle). This honeymoon period is followed within a few years by increasing numbers of negative news articles about the hot-selling new drug's dangers (stage 2). Suddenly everyone is alarmed that yesterday's wonder drug is today's looming threat. Then that, too, passes, and we get to stage 3, a more balanced attitude with a more sober understanding of what the drug can really do, as it settles into moderate sales and its proper place in the pantheon of drugs.

Then, *ta-da!*, a drugmaker releases its next miracle drug, and the cycle starts all over. When you hear the next breathless news report about the next breakthrough drug, remember the Seige cycle.

AS FOR THE TEN DRUGS I've chosen to highlight: You'll likely recognize some, while others will be new to you. The overall idea for this book came from my talented editor, Jamison Stoltz, but the final list is my own.

I didn't want to trot out the standard "greatest hits" list of drugs throughout history. So I've left out some of the usual suspects—aspirin and penicillin, for instance—because there's already been plenty written about them. In their place you'll find surprising chapters on lesser-known (but very important) drugs like chloral hydrate (knockout drops, used everywhere from doctors' offices to Mickey Finn's bar), and CPZ (the first antipsychotic, the drug that emptied the old mental asylums), along with a sprinkling of more famous drugs from the Pill to Oxycontin. The book includes a lot about opioids in their many forms, from the first prehistoric harvesting of poppy sap to today's murderously powerful synthetics. Opium's children are worth the attention because of their historical importance (their thousands of years of refinement and development illuminate much about the history of drug-making in general), their current importance (as agents of today's epidemic of addiction and overdose), and because their story is full of interesting characters and stories, from a genius medieval alchemist to a despairing Chinese empress to a lab littered with unconscious chemists.

Careful readers might notice that the number of drugs I highlight is more like ten-ish than exactly ten. Some chapters focus on a single chemical (like sulfa), others on a related chemical family of drugs (like statins). So don't get caught up in the count. That's not important.

What's important here is that nobody can pick the ultimate short list of history's most important drugs—it's futile to try—so I made choices based on my sense of the drug's historical importance plus its entertainment value. The writing style is designed to avoid as much scientific jargon as possible in favor of general readability; my preference is for lively stories and memorable characters. This might not make scientists happy. But I hope it works for you. Welcome to the world of drugs.

CHAPTER 1

THE JOY PLANT

YOU CAN IMAGINE an early hunter-gatherer in the Middle East looking for that next meal, roaming some new countryside, trying a taste of this or that insect, animal, or plant. Seeds, high in nutritional value, are generally worth trying. So, often, are the pods and fruits that surround them. On this particular day he or she finds a patch of waist-high plants growing in an open area, each head nodding under a heavy, fist-sized, waxy, light green seedpod.

Worth a try. A sniff. A small bite. A grimace and a spit. The flesh of the pod is mouth-twistingly bitter, and this is a bad sign. We are wired to sense a lot of poisonous things as bitter; this is nature's way of telling us what to avoid. Bitter usually means a stomachache or worse.

So our early explorer turned away from the plants with the big seedpods. Then an hour or two later, something strange. A dreaminess. An easing of pain. A pleasant sense of well-being. A connection with the gods. This plant was holy.

IT MIGHT HAVE started that way. Or it might have started when a sharp-eyed early human noticed some animal feeding on those same seedpods and afterward acting a bit odd, also a sign from the gods that the plant had power.

We do not know how it happened, exactly, but we know something about when. The long love affair between humans and this miraculous plant started more than ten thousand years ago—before towns, before agriculture, before science, before history. By the time the first human cities on earth were rising in the valleys of the Euphrates and Tigris Rivers, this holy plant's seeds were being eaten as food, its bitter sap was being used as a medicine, and its praises were being sung. During the excavation of a four-thousand-year-old palace in what is today's northwest Syria, archaeologists recently found an unusual room near the kitchens. There were eight hearths and a number of large pots, but there was no food residue. Instead, they found traces of poppy along with heliotrope, chamomile, and other herbs known to be used in medicines. Was this one of the world's first drug-manufacturing sites?

The plant at the center of this ancient attention was a particular strain of poppy. The seedpods, especially the sap in their outer walls, had effects that were so powerful, so healing, that it seemed almost supernatural. A terra-cotta statuette found on Crete and dated back more than three thousand years shows a goddess with a headdress adorned with pods of poppies, incised exactly as the pods are cut today to harvest the sap. "The goddess appears to be in a state of torpor induced by opium," wrote one Greek historian. "She is in ecstasy, pleasure being manifested on her face, doubtless caused by the beautiful visions aroused in her imagination by the action of the drug." Some archaeologists have proposed that the room in which this goddess was found was used by Minoans for inhaling the vapors of dried poppy sap.

The Greeks associated the plant with their gods for sleep (Hypnos), night (Nyx), and death (Thanatos), and put its image on coins, vases, jewelry, and tombstones. In myths, the goddess Demeter was said to have used poppies to soothe the pain of losing her kidnapped daughter, Persephone. The ancient poet Hesiod wrote eight centuries before Christ of a town near Corinth in Greece called Mekonê, which translates roughly as "Poppy Town," which some historians believe got its name from the extensive poppy farms that surrounded it. Homer mentions the plant in the *Iliad*, and in the *Odyssey* he tells the story of Helen making a sleeping potion, assumed by many to include poppy sap. Hippocrates mentioned poppy frequently as an ingredient in medicines. It was part of temple rituals, carved into statues, and painted on tomb walls. Dried and eaten or smoked, it was early man's strongest, most soothing medicine. Today it is among the most controversial. It is the most important drug humans have ever found.

IN A WAY it's amazing that early humans ever discovered any natural drugs at all. Consider that 95 percent of the three-hundred-thousand-odd plant species on earth are inedible by humans. Go out and start randomly munching the greens in your local woods, and the odds are twenty to one that you'll double over, throw up, or die. Among those few plants that are digestible, the chance of finding useful medicine is close to zero.

Yet our ancestors did it. Through trial and error, inspiration, and observation, prehistoric peoples around the world slowly found and built a store of herbal medicines. Early healers were locavores, using what grew close to home; in northern Europe effective herbs included mandrake root (for just about anything from stomach problems to coughs to sleeping problems), black hellebore (a strong laxative), henbane (to allay pain and ease sleep), and belladonna (for sleep and eye problems). Other early drugs, like cannabis, traveled on trading routes from points south and east. Many spices eagerly sought from traders in the Middle East and Asia, such as cinnamon and pepper, were used as medicines as much as seasonings. Early healers knew not only what their local herbs were but how to use them. A Greek physician in Nero's army in the first century, Pedanius Dioscorides, summarized what was known at the time in his multivolume *De Materia Medica*, one of the earliest and most important guides to drugs. In addition to listing hundreds of herbs and their effects, he described their preparation and recommended doses. Plant leaves could be dried, crushed, and added to potions brewed over slow fires; roots could be harvested, cleaned, smashed into pastes, or eaten fresh. Some could be mixed with wine, others with water. Medicines could be swallowed, drunk, inhaled, rubbed on the skin, or inserted as suppositories. Dioscorides's work guided the use of drugs in medicine for more than one thousand years.

He described the poppy, summarized its effects, and outlined its dangers: "A little of it," he wrote in *De Materia Medica*, "is a pain-easer, a sleep-causer, and a digester, helping coughs and abdominal cavity afflictions. Taken as a drink too often it hurts (making men lethargic) and it kills. It is helpful for aches, sprinkled on with rosaceum; and for pain in the ears dropped in them with oil of almonds, saffron, and myrrh. For inflammation of the eyes it is used with a roasted egg yolk and saffron, and for erysipela and wounds with vinegar; but for gout with women's milk and saffron. Put up with the finger as a suppository it causes sleep."

The plant and its magical juice accrued many names as it traveled from culture to culture, from the ancient Sumerian *hul gil* for "joy plant" to the Chinese *ya pian* (from which we derived the expression "having a yen" for a drug). The Greek word for juice is *opion*, which gives us today's word for the raw drug made from the poppy: *opium*.

You can't get it from just any poppy. There are twenty-eight species of poppy, members of the genus *Papaver*, on earth. Most of them are showy wildflowers that don't produce more than a trace of opium. Only two of the twenty-eight make appreciable amounts of the drug, and only one of these grows easily, suffers few pests, and doesn't require much irrigation. Its scientific name is *Papaver somniferum* (*somniferum* comes from Somnus, the Roman god of

Opium poppy (Papaver somniferum): white flowers, seed by M. A. Burnett. Wellcome Collection

sleep). This single plant, the opium poppy, still provides the world with almost all of its natural opium.

Researchers today debate whether this particular poppy was always so opium-rich, or whether early humans cultivated and bred it specifically to boost the amount of the drug. Whichever, by ten thousand years ago it was being grown in much the same way that it is today, and its medicine was being processed pretty much the same way.

Two thousand years ago, Dioscorides described how to gather the juice. It's remarkably simple: After a brief flowering, the poppy petals fall off. Within a few days the plant produces a waxy green seedpod that grows to the size of a hen's egg. Harvesters watch closely as the pod starts drying to a dull brown, and at the right moment they make a series of shallow cuts into its skin. These cuts weep the juice that contains the magic. The sap produced in the skin of the pod is where the drug is most concentrated (poppy seeds, used widely in baking and flavoring, contain very little opium).

Fresh poppy juice is watery, whitish, cloudy, and almost entirely inactive. But after exposure to the air for a few hours it turns into a brown, sticky residue that looks something like a cross between shoe polish and honey. That is when its medicinal powers are freed. It is scraped off the pod and formed into sticky little cakes, the cakes are boiled to remove impurities, and the resulting liquid is evaporated. The solid that's left, raw opium, is rolled into balls. And those dark, gummy balls changed history. Drugs before the nineteenth century were more than just bundles of herbs drying in the back rooms of witches, medicine men, and priests. They were processed and combined in ways part therapeutic, part magical—boiled into brews and elixirs, shaped into pills, mixed with everything from mummy dust and unicorn horn to powdered pearls and dried tigers' droppings, formed into elaborate concoctions for wealthy patients.

Opium was a prize ingredient. It could be dissolved in wine or blended into mixtures with other ingredients. It worked no matter how you took it—orally, nasally, rectally, smoked, drunk, or swallowed as a solid. One method might be a little faster than another, but no matter how it was delivered, it had the same range of effects, from making users sleepy and dreamy to killing their pain.

Most important—a sort of heavenly bonus—it made patients happy. It raised their spirits. It was more than a medicine; it was a doorway to pleasure. As one historian put it, "Opium was appealing because it always soothed the body while romancing the imagination. . . . Psychic and physical discomfort was replaced with hope and a halcyon calm." This was a truly seductive package of effects: a respite from pain, a feeling of well-being, a sense of exhilaration, an invitation to dream. Early users and caregivers often employed the same word to describe its effects: *euphoria.* Opium made it possible to bear the pain of disease and injury and at the same time to deeply rest. It was a perfect tool for early physicians (as long as it was used carefully; early healers, too, knew that too much could easily ferry patients from sleep to death).

It's no wonder the use of the drug spread across time through the Middle East and the Western World, from the Sumerians to the Assyrians to the Babylonians to the Egyptians, and from Egypt to Greece, Rome, and Western Europe. The best opium in the ancient world was said to come from the area around Thebes; one Egyptian medical text records its use in some seven hundred different medicines. The armies of Alexander the Great carried it with them as they conquered their way from Greece to Egypt to India, introducing it to local populations as they went. Poppy flowers became symbols of sleep both temporary and permanent, associated with the gods of slumber, dreams, and transformation, marking the passage from life to death.

The poppy's association with death was more than poetic. As early as the third century BCE, Greek physicians were already keenly aware that opium could be as dangerous as it was euphoric, and they debated whether the value of the medicine was worth the cost to patients. The Greeks worried about overdosing patients; they also realized that once patients started using opium, it was difficult to get them to stop. They wrote the first descriptions of addiction.

But the dangers of opium seemed far outweighed by its benefits. By the time Rome ruled the world in the first and second centuries CE, opium was said to be as widely consumed as wine and was sold on Roman streets in the form of poppy cakes—unbaked, malleable sweets made of opium, sugar, eggs, honey, flour, and fruit juice—used to lift the spirits and ease the minor aches and pains of the populace. Emperor Marcus Aurelius was said to take opium to help him sleep; the poet Ovid was also reputed to be a user.

After the Roman Empire's fall, opium found new markets thanks to Arab traders and merchants, who made the substance—lightweight, easy to transport, and worth its weight in gold for the right buyers—a standard part of caravan freight, spreading its use through India, China, and North Africa. One of the greatest of all Arab physicians, Ibn Sina (called Avicenna in the West), wrote around 1000 CE that opium is one of Allah's signal gifts for which he should be thanked every day. He very carefully described its many beneficial uses as well as its dangers, such as its ability to cause memory and reasoning problems, its constipating effects, and the dangers of overdose. Avicenna himself had seen a patient die from the rectal administration of too much opium. This great healer's thousand-year-old conclusion about opium sounds very much like the attitudes of today. "Physicians should be able to predict the duration and severity of pain and patient's tolerance and then weigh the risks and benefits of opium administration," he wrote, advising its use only as a last resort, and then recommending that physicians use as little as possible. It is likely that Avicenna was himself an early opium addict.

He and other Arab physicians worked it into cakes, infusions, poultices, plasters, suppositories, ointments, and liquids. Arab physicians of the Middle Ages were the world's best medicine makers, greatly expanding the art of drug-making by developing the use of filtration, distillation, sublimation, and crystallization, all part of a practice they called "al-chemie" (thought to be derived from the word *khem*, for Egypt, thus, roughly, "the Egyptian science"). The basic idea of alchemy, as it became known in the West, was to work with nature's raw materials to bring them to perfection, to help natural things evolve from their rough, raw states into more refined, more pure forms—to release their pure, inner spirits (this idea is embedded in our language: The alchemical distillation of wines and beers released the powerful liquors we still call "spirits"). Alchemy was

Avicenna Expounding Pharmacy to His Pupils. Wellcome Collection

at the same time a method of making useful items like medicines and perfumes, an exploration of the natural world, and an almost religious pursuit of the soul in all things.

Ancient Islamic writings made it clear that while opium could do great things, it could also enslave its users. Manuscripts also include descriptions of opium addicts, with their dangerous illusions, sluggishness, laziness, and diminished mental powers. "It turns a lion into a beetle," one writer warned, "makes a proud man a coward and a healthy man sick."

European use of opium declined after the fall of Rome, then grew again as soldiers trekking home from the Crusades brought the drug back with them from the Holy Land. By the sixteenth century it was being used from Italy to England to treat everything from ague, cholera, and hysteria to gout, itches, and toothaches.

Among its boosters was one of the strangest and most fascinating figures in medical history, a Swiss alchemist and revolutionary healer with the impressive name of Philippus Aureolus Theophrastus Bombastus von Hohenheim. Today he is better known as Paracelsus. He was a one-of-a-kind medical genius, part rebel, part con man, a bit mystical, a bit mad, a larger-than-life figure who trekked from town to town across Europe with his bags of remedies and instruments, carrying a huge sword with a pommel said to hold the Elixir of Life. He would come to a town, talk to the locals, hawk his skills, heal the sick, argue heretical new theories, pick up tips from local healers, and rail against the entrenched medicine of the day. "In my time there were no doctors who could cure a toothache, never mind severe diseases," he wrote. "I sought widely the certain and experienced knowledge of the art [of medicine]. I did not seek it from only learned doctors: I also enquired of shearers, barbers, wise men and women, exorcisers, alchemists, monks, the noblemen and the humble people." He listened, he argued, he learned, and he applied the best ideas to his patients.

Portrait of Paracelsus, whole-length.
Wellcome Collection

Along the way he penned several books, most of which were not published until after his death. These were written in a style that one historian called "very difficult to read and more difficult to understand," a mishmash of fantastic alchemical symbols and magical allusions, astrological references and Christian mysticism, medical recipes, divine inspirations, and philosophical ruminations. But underneath much of it lay a core of breakthrough ideas in medicine.

Paracelsus thought that most physicians were "vainglorious chatterers" who grew rich by simply parroting the mossy old ideas of the ancients, regurgitating the received wisdom of Roman and Greek and Arab authorities, repeating old mistakes. To this Paracelsus offered a simple alternative: True seekers of knowledge should read the book of nature. Instead of blindly following old texts from ancient authorities, physicians, he believed, should rely on what they see working in the real world, open themselves to all the wonders that nature offers, find new approaches, use new medicines in new ways, see what happens, and then use that knowledge to improve the art of healing.

Paracelsus experimented with his medicines, trying new mixtures and seeing what worked. (It's important to note that this was not experimentation in the modern scientific sense. It was more along the lines of "Here's something that looks interesting. I'll try it and see what happens.")

Chief among his successes was a mysterious and miraculous little black pill that seemed to ease almost any ill. "I possess a secret remedy I call laudanum and which is superior to all other heroic remedies," he wrote around 1530. One of his contemporaries remembered it this way: "He had pills which he called laudanum, which looked like pieces of mouse shit, but used them only in cases of extreme illness. He boasted he could, with these pills, wake up the dead, and certainly he proved this to be true, for patients who appeared dead suddenly arose."

Paracelsus's laudanum became the stuff of legend. We now know his secret recipe: About a quarter of each pill was raw opium; the rest was a fanciful (and mostly inactive) mix of henbane, bezoar stone (a solid mass gathered from the intestines of cows), amber, musk, crushed pearls and coral, various oils, bone from the heart of a stag, and, to top it off, a dash of unicorn horn (a much touted and certainly imaginary ingredient in many medieval medicines; what passed for "unicorn horn" in the day was most often the tusks of narwhals). Most of laudanum's effects came from opium.

Paracelsus was so sure of his views, so certain when he stated things like "The ignorant physicians are the servants of hell sent to torment the sick," or when he ostentatiously burned one of Avicenna's books in a public bonfire, that many considered him an arrogant braggart. But he was no charlatan. He was, instead, one of the fathers of pharmacology, a man who single-handedly helped wrestle drug studies away from the stranglehold of ancient theory and stand them on a more modern footing. He is said, for instance, to have studied opium by using it on himself and his followers, then tracking the effects—a practice of self-experimentation that would become common among physicians in coming centuries.

By the time Paracelsus died in 1541, the European appetite for opium was growing. Columbus had been briefed to look for and bring back opium from his voyages of discovery, as did explorers like John Cabot, Ferdinand Magellan, and Vasco da Gama. The reason was that opium, as opposed to many other Renaissance pills and potions, worked. As its popularity grew, so did the ways physicians found to use it. Some bright physician dissolved opium into a solution with mulberry and hemlock, then cooked the brew into a sea sponge. When this drug-infused "Sleepy Sponge" was dampened and heated, it released fumes that could both ease pain and put patients to sleep, making opium one of the first anesthetics. Venetian treacle, a mixture of opium with up to sixty-two other ingredients ranging from honey and saffron to viper's flesh, was used to treat everything from a snakebite to the plague. The popularity of treacle was so great that it helped spur the first drug regulations in London. In 1540, Henry VIII empowered physicians with the right to search apothecaries' shops and report any medicines found to be dangerous or defective, including treacle. By Shakespeare's day, only one man in London was allowed to make it, and even he had to show it to the College of Physicians before selling it.

One problem for early physicians using opium was that they never knew how strong the drug would be. Because opium came from different countries with different processing methods, there was no way to tell exactly what you were getting in a given ball. One pill maker's medicine might contain two or three or fifty times the dose of another's. Physicians had to try each new batch on their patients and hope for the best. Patients paid their money and took their chances.

The first steps to standardize the drug were taken in the 1600s by a renowned British doctor named Thomas Sydenham. Sydenham was a huge opium fan, a believer that this God-given substance was far superior in healing value to anything humans would ever be able to concoct on their own. He became famous for his own special tincture of opium dissolved in wine, with the bitterness of the drug offset by adding sweet port, cinnamon, and clove. Sydenham's liquid opium was easier for patients to swallow than pills. But the most important thing was that his preparation could be roughly standardized, the amount of opium in each bottle more carefully allocated, the doses more carefully measured. Sydenham made a fortune off this liquid form of opium, which he called—perhaps in honor of Paracelsus—"laudanum."

Portrait of Thomas Sydenham.
Wellcome Collection

Sydenham's laudanum was a hit, spurred by his own proselytizing; he so loudly sang its virtues that his friends nicknamed him "Dr. Opiophile." As sales grew, so did scientific interest in more precisely measuring its effects. British researchers like Christopher Wren and Gideon Harvey began experimenting with opium on cats and dogs, learning more about just how much was needed to achieve certain effects. They found new ways to test strength and ensure quality. Opium was helping to turn medicine from an art into a science.

It was also used for pleasure. One of the first books written in English specifically about the drug was *The Mysteries of Opium Reveal'd*, published in 1700 by the doctor John Jones. Jones told readers that the drug not only released one from anxiety but was also good for "Promptitude, Serenity, Alacrity, and Expediteness in Dispatching and Managing Business . . . Ovations of the Spirits, Courage, Contempt of Danger, and Magnanimity . . . Satisfaction, Acquiescence, Contentation, Equanimity," and so on. Opium raised feelings like "a most delicious and extraordinary refreshment of the spirits upon very good news, or any other great cause of joy." He compared its effects to a permanent orgasm. He sounded a lot like an opium addict.

The use of opium to alter mood rather than ease pain caught on at all levels of the social ladder. On March 23, 1773, the famed diarist James Boswell wrote, for instance, "I breakfasted with Dr. Johnson whose heaviness of spirits of yesterday was much relieved having taken opium the night before." The drug was being taken to ease depression.

Uses of all sorts rose along with a flood of new opium-containing medicines through the late 1700s, with names like Dover's Powder, Quaker Drops, and Dr. Bates' Pacific Pills. They could be purchased easily from doctors, at local pharmacies, even in grocery stores—no prescription necessary. Without laws to limit the use of these medicines, opium spread everywhere.

The public in Europe was eager for it. This was the era of the Industrial Revolution, when fast-growing populations of factory workers faced terrible working conditions. Underpaid workers living in growing slums needed a cheap release. Gin was one option, opium another.

Its popularity grew hand in hand with changes in disease patterns. Tuberculosis was one example: Fast-growing, close-packed urban industrial centers were breeding grounds for epidemic diseases like TB, a slow killer that often left victims in an agony that could only be relieved by opium. And there was cholera, carried in polluted water supplies and ragingly contagious, another disease that grew along with the slums. Cholera killed by causing uncontrollable diarrhea. Thankfully, one notable side effect of opium was its tendency to cause constipation; its use in cholera patients likely saved lives as well as soothed the dying. Growing numbers of prostitutes were among the drug's most faithful users, taking laudanum to ease the daily pains of their profession, to counteract the symptoms of venereal disease, and to lessen their despair. Sometimes they introduced their customers to the habit. Sometimes they used the drug to kill themselves. Doctors acted as sales agents for opium, promoting it to their patients and making a bit of money along the way. Chemists' and apothecaries' shops could count on opium-containing medicines to be among their best sellers and advertised accordingly.

And that was the thing about opium: Depending on how and when it was used, it could be a painkiller or a party drug, a lifesaver or a means of suicide. It was so popular in Western Europe by the end of the eighteenth century that some historians have linked it to the birth of the Romantic Era, with that period's emphasis on spontaneity, personal experience, relaxed morals, flights of fancy, and dreamy fantasies. It is certainly true that many of the leading artistic and political figures of the Romantic Age, from Byron and Berlioz to George IV and Napoleon, used the drug to one degree or another. Percy Shelley, drunk on opium, once burst into the rooms of Mary Wollstonecraft Godwin (with whom he was madly in love, although he was married to another woman at the time), a pistol in one hand and a bottle of laudanum in the other, declaring, "Death shall unite us." They lived long enough to marry; Mary's half sister, however, died of a laudanum overdose in 1814. Keats downed heroic doses. Samuel Taylor Coleridge and Thomas De Quincey were full-blown addicts.

"Nineteenth-century literature is steeped in laudanum," wrote one historian. And its appeal spread far beyond the intelligentsia. By midcentury, opium was as cheap as gin and more widely available in Britain than tobacco. Its use spread to the working classes, farmers, and the poor. Women took it for a break from the tedium of their lives, then gave it to their children to blunt their hunger and stop their crying. Men took it to ease their aches and forget their troubles. If there was some left over, they gave it to their farm animals to help fatten them for market.

One isolated and swampy rural area in England, the Fenlands, became infamous as the kingdom of the poppy. Malaria, with its recurrent fevers, was common there; so were rheumatism and the ague. Quinine (a malaria remedy from the bark of a South American tree) was too expensive for the local farmers. So were doctors. The poverty-stricken farmers turned to opium, not merely as a medicine but, as one observer noted, to "lift its user out of the mire of Fenland muck and the drudgery of agricultural life." A medical officer who visited the area in 1863 wrote, "A man may be seen occasionally asleep in a field leaning on his hoe. He starts when approached, and works vigorously for a while. A man who is setting about a hard job takes his pill as a preliminary, and many never take their beer without dropping a piece of opium into it."

It was considered a relatively harmless vice, certainly less dangerous than liquor. For every story about some baby inadvertently poisoned by too much opium-laced soothing syrup, there were others about long-term users doing just fine. Opium peddlers in the 1850s told the anecdote of an eighty-year-old woman who took a half ounce of laudanum daily for forty years without an ill effect. And didn't Florence Nightingale herself, the Lady with the Lamp, the very symbol of nursing care, occasionally use the drug? Of course she did. Would she do that if it was bad for you? Opium sales in Britain grew 4 to 8 percent each year between 1825 and 1850. To feed this growing national habit, the British encouraged poppy plantations in India, which soon became the source of much of the world's supply. The East India Company got into the business of shipping it around the world. Fortunes were made growing it, processing it, moving it, and selling it. And England was just the start. If opium was this popular at home, what might it be worth to traders if other countries were encouraged to use more of it?

A busy stacking room in the opium factory at Patna, India. Lithograph after W. S. Sherwill, c. 1850 by W. S. Sherwill. Wellcome Collection

India was one possibility. But the British needed their subjects in the Raj to have their wits about them. There were other targets, however: Countries where the opium trade could be expanded to Britain's benefit. Countries that might, in British eyes, be better off weakened by the drug. And so opium came to the most populous nation on earth: the Celestial Empire, China.

THE CHINESE ALREADY knew a thing or two about opium. They had learned about it first in ancient times, at least as far back as the third century. Arab traders had made it available, and Chinese alchemists had found it an interesting medicine. It was used in small amounts by the upper classes for treating dysentery and calming the concubines of wealthy men. For more than a thousand years there had been little more to it than that.

Then the first European sailors arrived. They wanted desperately to trade. They brought with them a number of items they thought the Chinese might value. But what did the Chinese need with scratchy British woolens or stiff Dutch linen when they had silk? What did they need with inferior Western pottery when they had porcelain?

There were a few things, however, that the Chinese did want. One was a pleasant new herb, the dried leaf of a plant from the Americas called "tobacco." The Chinese were fascinated by the sight of foreign sailors packing shreds of this leaf into small pipes and setting it on fire, breathing clouds of aromatic smoke. It had desirable effects. The Chinese elite quickly adopted the tobacco habit, and smoking became a seventeenth-century Chinese fad. The Europeans, happy to find something tradable, sold it in Canton by the shipload. If supplies ran low, the Chinese stretched their tobacco with other things, including adding shavings of opium and arsenic. The additions were thought to help stave off malaria. They certainly provided an extra kick.

Smoking became so popular in the Celestial Empire, and the addictive nature of the habit so obvious, that in 1632 the emperor felt it necessary to ban tobacco in all its forms. Just to make sure, he also ordered that all known tobacco addicts be executed. Tobacco disappeared. And during the ensuing drought, a few Chinese took to smoking opium alone.

There things stood until the early eighteenth century, when yet another valuable dried plant entered the picture. This one had long been grown in China, and, when steeped in boiling water, created a beverage with a pleasantly energizing effect. The British called it tea. And it quickly became as big a craze in England as tobacco had been in China.

As the English demand for tea grew, so did the need for merchants to find something—anything—the Chinese might take in trade for it. Tobacco was out. So British envoys were sent to the emperor's court with samples of tin, lead, cotton fabrics, mechanical watches, dried fish, anything that might appeal. Nothing did. "The Celestial Empire possesses all things in prolific abundance and lacks no product within its borders," the Chinese emperor sniffed around 1800. "There is therefore no need to import manufactures of outside barbarians in exchange for our products."

There might not have been an appetite for manufactured items, but there was one raw material the Chinese desired. Their currency was based on silver, and the Chinese had an endless hunger for the precious metal. This was bad news for the British, however, because most of the world's silver came from the Spanish holdings in the New World. The British had only so much silver, and the China tea trade was soon soaking that up, causing an imbalance in the world supply. Something else was desperately needed.

So attention turned to opium. Thanks to the spread of their opium plantations in India, the British had plenty of the drug to export. All they needed to do was turn the Chinese into users.

The Chinese emperors had no interest. Still smarting about the problems with tobacco, the Chinese government, in the face of British efforts to bring a new drug into their world, issued edict after edict restricting the trade in opium. The British found ways to keep bringing more in. Each new opium smoker was a new source of money, and once they started smoking it, they didn't want to stop. The lives of many Chinese peasants were as deadening as those of the Fenland farmers, and many became eager smokers. The rich and bored of China tried it as a lark, then bought more. The market grew. In 1729 the British sold two hundred chests filled with balls of Indian opium at the Canton port. By 1767 it had grown to one thousand chests. In 1790, four thousand chests. The Chinese emperors of the time—Hongli (the Qianlong Emperor) and his son Yongyan (the Jiaqing Emperor) were incensed. It was tobacco all

A fleet of opium clippers with other boats and rafts on the Ganges. Lithograph after W. S. Sherwill, c. 1850 by W. S. Sherwill. Wellcome Collection

over again—no, it was worse than tobacco. This new drug not only enticed, but it rendered its users indolent and unproductive. The emperors' edicts against opium got stronger, culminating in 1799 with a complete ban on the drug, a law barring all import of this odious and deplorable substance into the Celestial Empire. Officially, the British had to comply.

So they turned to smuggling. Before many years had passed, some twenty groups, from semi-legitimate businessmen to out-and-out pirates, were smuggling opium into China. These unscrupulous traders took over quiet little backwater ports along the Chinese coast, bribed local officials, and moved tons of Indian opium into China. The British government publicly deplored the scandal and privately looked the other way. The East India Company was heavily involved; a great deal of money was at stake. Certain activities were ignored, deals were made, and the opium kept moving from India to China, providing the money to move tea from China to England, and along the way helping a bit to destabilize the already shaky Chinese government. That, too, was good for the British. The weaker the government, the easier it would be to institute trade without the emperor's interference. Historians estimate that by the late 1830s about 1 percent of the entire Chinese population, some four million people, were addicts; near some smuggling ports, the proportion could be as high as 90 percent. By 1832, one-sixth of the entire gross national product of British India came from the opium trade.

Then the Chinese government decided to put a stop to it once and for all. The Opium Wars were about to start.

THE MATCH TOUCHED the fuse in 1839 when a sizable Chinese military detachment showed up outside the British trading post at Canton. The leader of the Chinese soldiers, speaking in the emperor's name, demanded that all opium sellers inside surrender their stocks of the drug. The commander of the small British military force looked over the Chinese soldiers massed outside and suggested to the traders that they comply. Thousands of chests of opium were handed over, and the Chinese promptly burned it in a huge bonfire in front of the British. They were making a point, both to the foreign traders and to their own people. Opium would not be tolerated.

Stung by the insult, Her Majesty's government (Queen Victoria had taken the crown just two years earlier) sent troops and warships to Canton, starting the first of two brief Opium Wars. The British won both handily. They weren't much as far as wars went: some minor skirmishes and small-scale naval clashes halfway around the world from Britain. But they did make some important points. First and foremost, the modern and well-equipped British forces, with powerful warships, smashed the outdated and outgunned Chinese military. The Chinese were faced with the fact that the Westerners had a superior military with better guns, better discipline, and far better ships. And opium itself played a role: By 1840 a large number of Chinese officers and soldiers were addicts, many too drugged to fight.

And second, the Opium Wars showed the Chinese that when it came to trade, the British would be calling the shots. When the battles were over, Britain reaped the spoils. The emperor ceded to Queen Victoria's government the port of Hong Kong as well as access to other ports and better terms for commerce. The Celestial Empire had been forced open.

But not to opium. Never to that. The British asked for specific government approval for the importation of opium, dangling the riches to be had from opium taxes. But even in his weakened position, the emperor of China drew a line. "It is true I cannot prevent the introduction of the flowing poison; gain-seeking and corrupt men will, for profit and sensuality, defeat my wishes," wrote Daoguang, the eighth Qing emperor, "but nothing will induce me to derive a revenue from the vice and misery of my people." He refused to legalize opium, a stubbornness on this point that arose, in part, from his personal family history. Three of the emperor's sons were addicts, and all three eventually died from the effects of the drug. Later it was said that Daoguang himself perished in 1850 from a broken heart. But until he was dead, he would never legitimize the opium trade.

It didn't matter. The drug was too well established. Hong Kong became the world's opium center, a vast drug marketplace where "almost every person possessing capital who is not connected with the government is employed in the opium trade," as the colony's British governor wrote in 1844. It was still technically illegal to move the drug into China, but as the smugglers' power grew, the British government turned a blind eye. Some of the opium runners grew into merchant princes, buying small fleets of opium clippers, the fastest ships in the world, to speed their cargo from India, and using their profits to buy baronial estates back in England. Large fleets of pirate junks, some controlled by smugglers, some preying on them, swarmed the coastal waters. China was falling into a lawless, malfunctioning anarchy. In the mid-nineteenth century, a combination of high taxes, starvation, and disgust with loose morals and the opium trade led to revolution—the Taiping Rebellion, led by a Chinese cult leader who believed himself the younger brother of Jesus Christ. It took fourteen years for the emperor to crush it. By then, more than twenty million Chinese had been killed and tens of millions more displaced. Many who were uprooted signed their bodies away as indentured workers—this was the start of what became known as the "coolie trade"—and left China forever.

As the Chinese state crumbled, and hunger and lawlessness gripped large parts of the empire, more of its people turned to opium. The *London Times* estimated in 1888 that 70 percent of the adult males in China were addicted or habituated to the drug.

And now it was spreading beyond China. Chinese coolies, shipped to America by the tens of thousands as cheap labor for mining, farming, and railroad construction, brought opium with them. By the 1880s, San Francisco was infamous for its twenty-six opium dens, iniquitous places that often featured gambling and prostitution behind the haze of smoke. Opium gained popularity with the city's demimonde, artists, bohemians, and wealthy white thrill seekers. This was the birth of America's drug subculture.

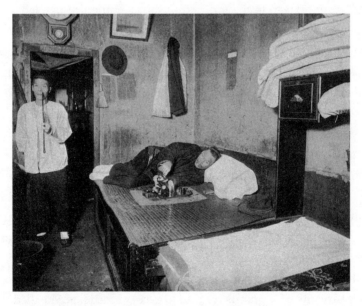

An opium den in San Francisco. Wellcome Collection

FINALLY, AFTER DECADES of profiting from the opium trade, even Great Britain had had enough. A string of sensational news stories in the late nineteenth century highlighting the corruption and tragedy in China disgusted the British elite and led to a

Parliamentary decision to end the trade. Almost all support, both official and unofficial, evaporated.

But the damage was done. Just before World War I, another Imperial decree went out, this one ordering that all opium smoking cease in China and that all dens close by 1917. The emperor by now was so weak, however, and the empire so powerless, that few users paid attention. That was true even in the Forbidden City, where the wealthy elite, exempted from the drug edicts that affected the rest of the country, continued to smoke the drug.

Which leads to the story of Wanrong, the wife of the last Chinese emperor. This beautiful young woman, born in 1906 and married at age sixteen to the indifferent young emperor Puyi, led a life that was pampered, purposeless, and almost completely devoid of love. At an early age she started smoking opium. And she never broke the habit. For decades, through the final decline of Imperial China, through the revolutions and invasions of the 1920s and 1930s, through World War II and eventual abandonment by her husband, she took ever-increasing solace in the drug. By 1946 the empire was dust, and Wanrong was a prisoner both of her habit and of the Chinese Communists.

They made a show of her. They threw the empress into a cell, humiliated her, and kept her from her drug. Soldiers and peasants were allowed to file by and look in through the bars, laughing and goggling. Wanrong went into severe withdrawal, her rags spattered with vomit and feces—muttering, weeping, crying orders to imaginary servants. Her guards refused to clean or feed her. She died of a combination of malnutrition and withdrawal in 1946.

This was the new reality in China. In 1950, the Communist government outlawed the cultivation, sale, and use of all narcotics. After the British left the trade, the Chinese had started growing their own poppies. Now those poppy fields were burned, plowed under, and turned to food production. Stocks of opium were burned. Dens were torn down. Tens of thousands of dealers and addicts were jailed, reeducated, and, if they persisted in their use, killed.

That is what it took to break the nation's long addiction. By 1960, opium had finally been wiped out in China.

But the drug was too powerful, too seductive to die.

ON A TRIP to Paris in the late 1700s, Thomas Jefferson was introduced to La Brune, an oily, dark French medicinal concoction whose major attribute was its significant jolt of opium. Jefferson was so impressed that he brought some back with him, recommending it to his friends in the new United States as a capital remedy for all aches and pains.

It was the start of a craze. Then as now, "Americans always want to try new things," as one publication of the day declared, from a new mechanical device to a new patent medicine to a new drug. The new Republic had plenty of small drug firms that eagerly started making opium-laced elixirs, extracts, and tonics. Many of them were easy-to-take liquid variations on Sydenham's laudanum.

The nineteenth century was the age of patent medicine in America, of mass advertising and medicine shows, snake-oil salesmen and wild claims, a time in which the nation was wide open to the over-the-counter sales of just about any drug anybody would pay for. Patent medicines—so-called not because they were patented in today's sense, but because back in England certain nostrums used by the royal family were granted "letters patent," allowing makers to use a royal endorsement in advertising—were, by the mid-1800s, a huge business in America. Pushed by some of the first efforts in mass advertising, sales of these over-the-counter brews were powered by laughably inflated claims, high alcohol contents, and, often, opium. Corner drugstores offered cures like Stott's Unique Fruit Cordial (unique because it contained 3 percent opium), Mrs. Winslow's

Soothing Syrup (opium in a sweetened form ideal for fretful babies), and Chlorodyne (a mixture of laudanum, cannabis, and chloroform). Doctors recommended opium nostrums to patients with rheumatism, cholera, and just about anything that caused physical discomfort, from childbirth to gout. Opium-laced patent medicines might not be able to cure cancer (as some makers claimed), but they certainly eased pain, soothed coughs, and lifted spirits. Opium use in the United States skyrocketed, with imports of the drug rising from 16,000 kilograms in 1840 to 44,000 ten years later, and 250,000 kilos by 1870.

Along with increasing use came increasing risk. Accidental overdoses became more and more common among children. And not all of them were accidental. There were occasional reports of parents using an overdose of a soothing syrup to rid themselves of an unwanted baby. Child welfare agencies and charities began to raise the alarm.

In adults, the problem was addiction. As early as 1840, public concern began to focus on those who could not get off the drug, pointing to cases like Edgar Allan Poe's wife, who, dying from tuberculosis, tamed her pain with what one historian described as "staggering" doses of opium. Poe himself was rumored to be a user, perhaps an addict. He was one of thousands.

Many physicians continued to recommend the drug to their patients. Addiction, in the context of mid-1800s America, was not seen as such a terrible thing. Even those doctors who found the use of opium deplorable believed for the most part that if the drug was properly controlled by the patient and overseen medically, it was a fairly benign habit. In any case it was certainly better than alcoholism.

Drinking was America's particular curse. Drunks were loud, wild, and sometimes violent—they shot guns and got into fights—while opium users were peaceable, reserved, and often surprisingly happy. "Liquor generally arouses the animal," a correspondent wrote in the *New York Times* in 1840, "while opium subdues this completely. Indeed, in its place it awakens the diviner part of human nature and can bring into full activity all the nobler emotions of the human heart." Most doctors saw opium addiction as a private matter, an unfortunate personal weakness to be treated with sympathy, the patient slowly eased off their habit and, if necessary, supplied with maintenance doses of the drug for as long as needed. After all, many, perhaps most, addicts had been launched on their habits by doctors intent on easing their pain during medical treatment for illness or injury. Even when hooked, opium tipplers remained more or less functional as long as they got their minimal dose. It wasn't so bad.

Then modern science intervened, and the scene changed dramatically.

OPIUM WAS FASCINATING to researchers as well as users. The old alchemists had long given way to modern chemists, their powers vastly increased by the use of ever-more-powerful scientific techniques and equipment. But some things hadn't changed that much. Modern chemists, like the old alchemists before them, were still interested in pulling natural substances apart, finding out what made them tick, purifying the pieces, and recombining them in new ways. Chemists wanted to know what central component gave opium its power. Physicians wanted purer, more refined, more standardized preparations of opium for their patients. They all wanted to get to the heart of the drug, to find and work with the specific chemical that gave it its healing power, its euphoric kick.

The first breakthrough came in 1806 when, out of nowhere, a young German apprentice pharmacist named Friedrich Sertürner, working by himself in a crude laboratory, discovered opium's soul. He spent months finding ways to gently heat, dissolve, and pull apart the raw, sticky stuff, to separate it into its pieces, and to purify them using different solvents and methods of distillation, cooling vapors into liquids, drying liquids into crystals, and redissolving the crystals in new solvents. By doing this he created hundreds of new preparations. He tested what he made on stray dogs, then on a few of his friends, and finally on himself.

Sertürner found that opium was not one thing but many, a complex cocktail of ingredients. The most powerful of these were members of a chemical family called alkaloids—all of which shared some common molecular structures and attributes, and all of which were bitter to the taste. It turned out there were three or four major alkaloids in opium, and possibly dozens of minor ones.

Sertürner was the first to roughly purify and study the most important of these, the alkaloid that gave opium most of its power. Separated from the natural mix, this chemical by weight had effects ten times more powerful than opium. He first called his substance the *principium somniferum*, the central sleep-making principle, for its ability to put people into a drowsy stupor. Later he named it "morphium," after Morpheus, the Greek god of dreams. Today we call it "morphine."

It was an astounding achievement for an unknown amateur chemist in his early twenties. Perhaps because of that, it was roundly ignored at the time. Sertürner was a nobody, and few serious scientists paid his work much heed. The young man kept at it though, getting purer and purer versions of his morphium, taking dose after dose to test them, carefully noting how his moods changed.

It started out so beautifully for him, with hours of euphoria, soaring dreams, the end of pain. Then he began to suffer from constipation. When he tried to stop the drug, there was deep depression and a gnawing hunger that almost drove him mad. He went back to it and tried increasing his doses. On one occasion he nearly killed himself and three of his friends by ingesting, at half-hour intervals, enormous amounts of morphium; their lives were saved only at the last moment when Sertürner, marshaling the last of his wits, gave them all a chemical to induce vomiting. It turned ugly. By 1812, after years of research, he was horrified by what he had done. "I consider it my duty to attract attention to the terrible effects of this new substance I call morphium," he wrote, "in order that calamity may be averted."

Sertürner lived until 1841, starting his own pharmacy, making a decent living, and dying in obscurity. He never made a fortune off his morphium.

That was left to others. The study of alkaloids took off after Sertürner's work, and in the 1820s other, better-known scientists began working seriously with morphine. One old German pharmaceutical firm became expert at making it in bulk. You might have heard of Merck. It makes many drugs now, but morphine was the rock upon which its empire was built.

The ability to pull apart raw substances, to purify and study their active ingredients, fueled the new science of organic chemistry, the study of the molecules of life. Organic chemistry and drug-making grew up together. Through the nineteenth century, other researchers picked apart more of the opium cocktail, purifying the other alkaloids in the raw drug. There were a lot of them. Codeine, isolated in 1832, was less effective at killing pain than morphine, but also less likely to cause addiction; we know it now mostly for its use in cough syrups. Then thebaine, noscapine, papaverine, narcotine, narceine—the list of opiate alkaloids lengthened. As alkaloid chemists became more skilled, more alkaloids—cocaine, nicotine, caffeine, strychnine, quinine, atropine—were isolated from other plants like coca, tobacco, coffee, nux vomica, and the bark of the cinchona tree. The list of alkaloids grew. They were all chemically related, all active in the body, and all bitter.

But morphine was the first and most important. It quickly replaced opium in medical use. It could be made to exact standards and strengths, making possible more accurate dosing and giving physicians a better tool with which to treat their patients. It was a far stronger painkiller than opium, becoming a staple medicine in hospital pharmacies and doctors' bags. Its only drawback was that in the early days it had to be taken either by mouth or via wax-coated suppositories, which slowed its action and made results more variable. Even after drinking a liquid form, patients had to wait for the drug to hit, then the effects built gradually, making it harder to adjust doses.

Physicians wanted a better way to get morphine into the body. They tried powdering it and having patients inhale it, but that was too likely to cause nausea. They tried smearing it on the skin, but it raised blisters. They tried inserting it under the skin, using splinters or needles to force little balls of medicine into small incisions, but it was too difficult to control the dose.

The answer came in 1841 when a French surgeon named Charles Gabriel Pravaz introduced a new tool to medicine. Pravaz was looking for a way to treat varicose veins and thought that using drugs to slow blood clotting might help. The trouble was that the drugs he wanted to use, when taken by mouth, were destroyed in the stomach. He needed a way to deliver them directly into veins. So he asked a local metalworker to craft a hollow needle out of platinum, and to the needle he attached a small silver plunger. The idea was to load the plunger with the drug, insert the needle into the vein, then push in the drug.

He had made the first syringe. With it, Pravaz could draw in a precisely measured amount of a drug and deliver it through the skin, directly into the body, bypassing the vagaries of the stomach and gut, speeding its action, getting more of the drug where it needed to go. Pravaz carried his in a silk-lined pocket sewn into his top hat. And his invention, the "Pravaz," as it was often called, quickly became a hit among doctors. It gave physicians a vital new way to deliver drugs more quickly and more accurately.

The Pravaz was perfect for morphine. Shooting the drug directly into the body could transform agony into calm within moments. Nurses, faced with a patient writhing in pain, could pull out a syringe of morphine and say, as one anecdote went, "I'm about to become your best friend." Doctors were able to run more accurate studies.

The new, purified drug also offered hope to opium addicts. The idea among some doctors was that by treating addicts with lower, more measured doses of morphine, they could blunt the appetite for opium and wean their patients off it.

Of course it didn't work. Morphine was basically the same drug as opium, only stronger. At best, it was a replacement for opium, not a cure. Shooting morphine with a Pravaz made it easier for addicts to get a bigger, faster rush. The danger of addiction rose accordingly.

By the time of the American Civil War in the 1860s, morphine was a battlefield staple, shot into soldiers to ease the pain of wounds and to treat the dysentery and malaria that raged through military camps. Home gardens in both the North and South were ablaze with poppy flowers as citizens patriotically grew opium for their troops, the raw drug processed into morphine and rushed to the front. Millions of doses were given. Thousands of veterans with lifelong wounds—limbs torn away, bones crushed, spirits broken—were taught how to use syringes to self-administer the drug long after the war ended.

The result was a wave of addiction they called "the army disease." Thanks to morphine, per capita use of opiates tripled in the 1870s and 1880s, creating America's first opiate crisis. Anyone could get morphine and a syringe to shoot it; they were sold by mail order and over the counter at drugstores. As morphine's medical use increased—for surgery, for accidents, for pretty much any disease or injury—so did the number of patients dependent on the drug.

Scientists called this new epidemic "morphinism" and tried with increasing concern to find ways to control it.

The opiate crisis of the 1880s sounds a lot like the opiate crisis of today, not only in the way the number of users was soaring, but also in the ways that society responded. At first physicians and government officials tried "soft" approaches, minimizing the problem as less serious than, say, alcoholism; moderating their recommendations for the drug, looking for better ways to ease patients off of it; even experimenting with municipal narcotic clinics where opiate addicts could get maintenance doses of their drugs. Druggists, too, took note. While opiates were an important revenue stream for many pharmacies, others decided not to sell the drugs at all. "A greedy criminal druggist will sell you morphine or cocaine," a sign read on one New York pharmacy, adding, "We are not of that kind."

But there are differences, too. Today's opiate addicts are sometimes considered lower-class, either big-city junkies or rural white trash. But in the 1880s, morphine addicts (apart from veterans) were mostly members of the middle and upper classes, professionals and businesspeople who had once been in pain and whose physicians had taught them how to self-inject the drug. Physicians themselves were among the most dedicated morphine users; by one 1885 estimate, up to a third of the physicians in New York City were addicts.

In many ways morphine was a woman's drug, recommended for the treatment of a variety of women's problems, from menstrual cramps and hysteria (which at the time was a catchall term for just about any psychological problem suffered by a woman), to depression (or, in the terms of the day, melancholia). It's striking that throughout the 1800s, the majority of laudanum and morphine users in the United States were women. Alcohol and tobacco were considered men's drugs; for women, opiates became the avenue of escape from lives severely restricted by social norms and standards of etiquette. Many women who started with medically recommended laudanum or morphine became addicts, indulging in a quiet, private, easy-to-hide habit, an open secret in many professional households. Morphine replaced laudanum for many of the era's upper-class invalids, those aging maiden aunts and gouty grandmothers who would retire to their rooms complaining of exhaustion or "nerves" and take comfort in their Pravaz. As one historian noted, in the 1870s "a typical Southern addict was female, Caucasian, reasonably well off, and addicted through medical use." Just before World War I there was even a short-lived fad for treating women in labor that doctors called "Twilight Sleep." Physicians dosed women in labor with a combination of morphine and an anti-motion-sickness drug, promising them a pain-free delivery. Later it turned out that the treatment didn't stop the pain so much as it erased all memory of it. Some women screamed so much during Twilight Sleep that they had to be put into soundproof rooms. But when they awoke, baby in arms, they thanked their doctors. They'd forgotten the experience. Twilight Sleep Associations sprang up in major cities.

Medical treatment often started the morphine habit, but medicine was limited in what it could do to help patients kick it. Doctors, increasingly concerned about morphinism around the turn of the twentieth century, gently encouraged their patients to gradually lower their doses. Other than that, they were unable to do much.

The concept of addiction, both physical and psychological, was not well understood, the mechanisms were unknown, and cures were often left up to the patient. Most addicts had money; if they wanted to get off the drug, they could afford a stay in one of the many private treatment centers and sanatoriums that were popping up in major cities—the start of what today we call the drug rehab business. Here they could take a break from their habit. But there was little to stop them from starting up again.

For drugmakers, both morphine and morphine cures were ways to make money. Drugs and drug-making were wide-open businesses, free of almost all legal oversight. Just about anybody could gin up an over-the-counter remedy promising to cure anything and everything, including morphinism. Many of these cures were useless mishmashes of mild herbs mixed with heroic doses of alcohol. Others contained opium or morphine itself, a cure that simply extended the disease.

Morphine made the old problems with opium look quaint. Back in the Romantic Era, laudanum drinkers usually started out drinking around one fluid ounce of the medicine a day (about half a shot glass of most preparations). That amount contained opium equal to about one grain of morphine. Serious laudanum addicts might work themselves up to five or six of these tipples per day—about six grains of morphine. In contrast, a seasoned morphine addict using a syringe in the 1880s was shooting up as much as forty grains a day.

Doses like that could kill a beginner. And that was another problem. Morphine could be a killer. It is a drug with what's called a narrow therapeutic window—a small range of doses in which it works. Too little, and the pain remains unbearable. Too much, and the patient stops breathing. Because the dose you need is very close to the dose that will kill you, it's easy to overdose. And that happened increasingly to morphine users in the years leading up to 1900.

By the late nineteenth century, morphine was by some estimates the most popular method of suicide for women and second only to guns among men. For decades it was also a popular means of killing others—overdosing a victim with morphine was easy, cheap, and virtually undetectable (the first good test for morphine in the blood or urine wasn't developed until the 1930s). By 1860, opium and morphine were suspected to be responsible for one-third of all poisonings in the United States.

Tragic morphine stories like this were newspaper staples: In the 1890s, the teenage daughter of Eberhard Sacher, a respected

Viennese professor and an expert on women's diseases, became pregnant out of wedlock. The girl suffered through a botched abortion that left her in excruciating pain. So her father treated her with morphine, and she became addicted. He blamed himself. What happened next is unclear, but no less poignant for that. Trapped between the scandal, his daughter's pain, and his own despair, in 1891 Sacher went to his medical supplies and pulled out a needle. Hours later, both he and his daughter were dead of morphine overdoses. Perhaps her death was an accident, perhaps it was a planned murder-suicide, there is no way of knowing now. The news shocked Vienna and ignited calls for morphine regulation in the Habsburg Empire. But nothing official was ever done. Little, it seemed, *could* be done.

By the time the nineteenth century turned into the twentieth, however, inaction was no longer an option. There were too many suicides, too many accidents, too many murders, too many lives lost to addiction. Something had to be done. There had to be something—a new drug, a new wonder from the laboratories—that could undo all the damage. So scientists threw themselves into a search for a more benign drug, something that would still ease pain, but without the danger of addiction and death. It was the start of a century-long scientific search for safer, nonaddictive opiates.

A second effort was legal. Government officials woke up to the fact that opiates had to be controlled. The result was a storm of regulations, wars on drugs, the demonization and criminalization of both drugs and their users, and one hundred years of attempts to crush the problem through government action.

IF I HAD to pick one drug, above all others, that influenced the intertwined histories of medicine and pharmaceuticals, it would be opium. It's not merely because of the power of the drug and its deep roots in history. It's because the drug illustrates, more vividly and directly than any other, the dual nature of drugs in general: their power to do great good on one hand and great harm on the other.

You don't get the good without the bad. Every scientific discovery is a two-edged sword, with benefits inevitably tied to dangers both physical and psychological. Often humans jump at the benefits and leave the dangers to be dealt with later. That's certainly what happened with the joy plant, God's own medicine, opium.

CHAPTER 2

LADY MARY'S MONSTER

MARY PIERREPONT was strong-willed, pretty, and enamored of books. She was doubly lucky from the start: not only born into the English nobility in the late 1600s and therefore rich, but also a member of a family devoted as much to learning as it was to status. Her great-grandfather had helped found the world's first scientific organization, the Royal Society, just three decades before she arrived on the scene in 1689. Their main family home was graced with one of the world's largest and finest private libraries. Her father served in Parliament. Hers was a charmed childhood, lived in elegant houses, with the best food, the wittiest visitors, and opportunities for education far beyond most women of her time. And Mary, growing into a lovely woman known for her fine eyes and auspicious marriage prospects, thrived in it. She was smart; she knew it; and her family cultivated her intelligence. As a teenager she read her way through the family library, taught herself Latin, wrote poetry, and corresponded with bishops.

But she wanted more. She was determined to become that rarity of rarities: a female writer. She could not abide being told what to do and treasured her independence, so when her father tried to arrange a marriage against her wishes, she broke with her carefully picked groom and eloped with her own choice: Edward Wortley Montagu, grandson of the Earl of Sandwich. Their scandalous union was a hot topic among the social set for a season. But, then, it could have been worse. Montagu was, at least, from a fine family. And he had ambitions to rise in the government.

Mary began to publish some of her writing and earned a bit of notice for a few poems. Her wit could be biting: Some of her poems were so barbed, so targeted at members of her social set, that she decided to distribute them anonymously. She was earning a reputation as one of the brightest women of the age; Montagu was working his way up the political ladder. They had their first child, a son, in 1713. Their lives seemed charmed.

Then the Speckled Monster struck.

It took her brother first. He was only twenty years old and a particular favorite of Mary's. The disease hit him suddenly, put him to bed in an agony of pain and fever, and disfigured him horribly. He was dead within weeks.

The disease was called "the small-pox" (to distinguish it from the great pox, or syphilis). It was a fact of life in England, and through most of the world in those days, the biggest killer of the time, spread through wildfire epidemics afflicting millions, preferring to kill the young rather than the old. In the first day or two of its appearance it could have been mistaken for the common flu, little more than a headache and a mild fever. Then it got worse, causing a racing pulse, a fever high enough to make the patient sweat, constipation, vomiting, and an unslakable thirst. After a few days, an itchy rash of small pink spots rose on the skin, growing darker and burrowing deeper, developing into foul-smelling pustules that begged for scratching. Sometimes there were just a few dozen spread across the chest and back. Other times there were thousands, the patient's skin—lips, mouth, throat, nostrils, eyes, and sexual organs as well—turned into a carpet of pustules, a fiery, blistered, itching agony. The body reacted to this assault with increasing fever. Patients might swell up, their skin ballooning and stretching so much that faces were sometimes unrecognizable. Noses and throats could swell shut, breathing turning to gasps as the airways closed. The pustules filled and became tender, bursting against the bedclothes, releasing a thick, stinking, yellow pus. Rest was impossible.

Some physicians thought the best treatment was to sweat out the poison, so they piled on blankets and built up the fire. It didn't work. Others went the opposite way, wrapping the patient in cold, wet sheets and throwing open the windows. That didn't work, either. Nor did bleedings, purgings, laxatives, induced vomiting, or any other of the standard medical treatments of the day. Nothing helped.

No one knew what to do, because in the early 1700s they didn't know what caused the disease. In the end, all they could do was try to ease the discomfort, support the worried family, and wait. Within a few days of the pustules appearing, one of two things would happen. In about a quarter of the cases the disease progressed and the patients died. But the rest of the time, the patients would rally and throw off the disease, their fevers breaking, their pustules beginning to dry and flake. After some days or weeks of recovery, they could totter out of their sickrooms and rejoin the world.

Alive, but marked. Smallpox blinded some of its victims and disfigured many of them. Almost every survivor was scarred by deep, disfiguring pits in their skin where the pustules had been, "turning the babe into a changeling at which the mother shuddered, making the eyes and cheeks of the betrothed maiden objects of horror to the lover," as one observer of the day wrote. Most adults in Britain carried these scars. It is said that the style of wearing veils, heavy makeup, and false beauty marks arose as ways to hide the effects of the disease. For a while, women's fashion was to paste small bits of fabric cut into crosses and stars over the worst of it.

And so it had been for centuries. Smallpox was fiercely contagious; today we know that you could get it from breathing in a bit of flaking skin, touching the pustules of a patient, or just handling their clothes. In Mary Montagu's day, the appearance of smallpox in town meant that you were better off leaving for your country home. Unlike other killer diseases of that time (such as cholera, which was generally limited to the poorer parts of a city), smallpox did not distinguish between rich and poor. It raged through palace and slum alike, killing kings as easily as commoners. It remains the all-time champion of contagious diseases, the greatest infectious killer humans have ever experienced. In Europe it killed more victims than the Black Plague, "filling the churchyards with corpses," a 1694 observer wrote, "tormenting with constant fears all whom it has not yet stricken, leaving on those whose lives it spared the hideous traces of its power." When European explorers and conquerors carried it with them to lands that had never seen smallpox, the result was a holocaust. It wiped out entire tribes in Africa, killed most of the Aztecs and Incas in the Americas, then spread along with the Europeans, decimating most of the tribes in North America, a sort of biological genocide that cleared the way for white pioneers. In Lady Mary's time it was just beginning to devastate the Aborigines in Australia.

The only good news—if you could call it good news—was that if you lived through the disease, you never got it again. This was a blessing of a sort: Smallpox survivors could safely nurse the afflicted with little fear of getting the disease. But no one knew why this was true, either; it was just another mystery in an age of mysteries. These matters of disease and life and death were almost entirely beyond human understanding. Only God could send disease, and only God could determine its result. Only God could winnow the living from the dead.

Here's the remarkable thing: Today there is no more smallpox. There has not been a single case on earth since the 1970s. Between Lady Mary's time and our own, we somehow succeeded in wiping humanity's worst disease enemy from the globe. This is perhaps the greatest success story in all of medicine. And it started with Mary.

TWO YEARS AFTER her brother's tragic death, Lady Mary Wortley Montagu—now living in London with her fast-rising husband—came down with a fever. Then spots appeared. Her physicians had little doubt about what she had. She took to her bed, another victim of smallpox, and the disease marched through its stages. The physicians were not sanguine—hers was a serious case. The pox spread and deepened; she tossed and scratched. The doctors told her husband to prepare for the worst.

But Mary was fated for other things. She made it through the crisis and threw off the pox. Weeks later, she opened her bedroom door and showed herself. Her eyelashes were gone. The skin around her fine eyes was red and irritated, and it would remain that way for the rest of her life, giving her a somewhat fierce look. The once-smooth skin of her face was marred by pits and scars. But she was not blinded, like so many other victims. And her spirit seemed intact.

Soon after, her husband was named His Majesty's ambassador to the Ottoman Empire, a fine promotion, and was ordered to Constantinople (now Istanbul) to take up his duties. Montagu expected to go alone; given the rigors of long-distance travel in 1715, it would have been conventional to leave his wife and child at home during his sojourn abroad. But Lady Mary was anything but conventional. Her strength had returned; her curiosity about this

Lady Mary Wortley Montagu.
Lithograph by A. Devéria after C. F.
Zincke. Wellcome Collection

strange foreign land was keen; she would not miss this adventure. She insisted on coming with him, bringing their young son as well.

So began a months-long trek across Europe and into the exotic lands of the East. Along the way she wrote a series of remarkable letters describing the regions they traveled through. Lady Montagu was more frank and highly observant—and less prejudiced against foreign customs—than most writers of her day; when published later, her letters became early classics of travel literature. This, too, might have been part of her plan: The journey to the Ottoman Empire offered a chance to build her reputation as a writer.

Once installed in the European section of Constantinople, with her husband away all day at the embassy, Lady Mary began learning everything she could about this strange Muslim world. She was especially interested in the lives of women. Europeans in general viewed the Ottomans as barbaric throwbacks who kept slaves and imprisoned their women in harems, beheaded nonbelievers and wailed their religion all day from the tops of towers. It was as though the Ottomans still lived in the Middle Ages.

Lady Mary came to believe otherwise. Her position as the ambassador's wife opened the door to friendship with some of the leading women of the city, elegant noblewomen who offered her unprecedented access to their apartments, their baths, their food, their customs, and their thoughts. She came to understand that the Ottoman system—with women living in all-female seraglios, separated during worship, and denied direct action in politics—was seen by the women as less an imprisonment than a path to a peculiar sort of freedom. Her new friends did not seem bullied or deprived; they were cultured, intelligent, seemingly very happy, and empowered in ways she hadn't imagined. Yes, they spent a lot of time among other women, but within that world they were freer than many European women, free with their opinions, free to express themselves. They were intelligent and well-informed. They had strong female friendships based on simple affection. She came to see them as experts in exerting power indirectly. These were women leading full lives—if very different lives—from modern European women, who all too often spent their time competing with other women for power and attention in a world of men.

And they were free with their bodies. They were amazed at the armor Lady Mary wore, her heavy gowns and stiff stays and corsets; she was amazed at the casual nudity of their baths. One of the many small things that caught her attention was the unflawed, beautiful skin of the Muslim women. Where were their smallpox scars?

She found out, and wrote about it in a 1717 letter: "I am going to tell you a thing, that will make you wish yourself here. The small-pox, so fatal, and so general amongst us, is here entirely harmless, by the invention of engrafting, which is the term they give it. There is a set of old women, who make it their business to perform the operation, every autumn, in the month of September, when the great heat is abated. People send to one another to know if any of their family has a mind to have the small-pox; they make parties for this purpose, and when they are met (commonly fifteen or sixteen together) the old woman comes with a nut-shell full of the matter of the best sort of small-pox, and asks what vein you please to have opened. She immediately rips open that you offer her, with a large needle (which gives you no more pain than a common scratch) and puts into the vein as much matter as can lie upon the head of her needle, and after that, binds up the little wound with a hollow bit of shell. . . . The children or young patients play together all the rest of the day, and are in perfect health to the eighth. Then the fever begins to seize them, and they keep to their beds two days, very seldom three. They have very rarely above twenty or thirty in their faces, which never mark, and in eight days' time they are as well as before their illness. . . . There is no example of any one that has died in it, and you may believe that I am well-satisfied of the safety of this experiment. . . ."

This "engrafting" was one of the first Western descriptions of what today we call inoculation. Lady Mary's description of the technique was accurate except in her use of the word *vein*, perhaps an indicator of her lack of medical knowledge. The Turkish technique was to make a simple scratch, usually on the arm, just deep enough to cause bleeding. Into this was placed a needle-tip's worth of a mix of powdered scabs and/or pus from the pox of a patient with a mild case; this "smallpox matter" then kicked off a mild case of the disease. After it passed, the child no longer had to worry about getting smallpox.

Lady Mary was fascinated. She probably discussed the procedure with the physician to the British embassy, and spoke with the French ambassador, who assured her that the practice was as common and harmless as taking the waters (going to a spa) would be in Europe. A few European physicians had already described the practice in positive terms in letters home, but with no effect on medical practice. So she began thinking about doing something very brave, and perhaps very foolish: She considered having this "barbarian" engrafting done to her own son.

She had to move fast: Her husband had been informed that he was being recalled to England. So, without his knowledge, Lady Montagu arranged to meet an old woman skilled in the technique and talked the embassy's surgeon—a somewhat reluctant Scot named Charles Maitland—into joining her and observing. The old woman arrived, armed with matter from a fresh blister of a suitably mild local case, and took out a long (Maitland noted, rusty) needle, scratched the boy's arm deeply enough to make the six-year-old yowl, mixed some of the matter with the boy's blood, and rubbed it into his wound. Maitland then jumped in. To ensure results, engrafting was often done on two arms, and Maitland decided to save the boy more pain from the needle scratches by using his surgeon's scalpel to score the other arm. He himself put in a bit of the pox, then bound the wounds.

And they waited. As hoped, a week later the boy developed a mild case of smallpox and then recovered, completely, without any scarring. Lady Mary had protected her son. He would never get smallpox again.

This was the crucial point: In Turkey, Lady Mary and Maitland learned how to purposefully cause a mild case of smallpox in a child to prevent a much more serious—and perhaps fatal—case later. This was personal to Lady Montagu: Had her brother been similarly inoculated, he would still be alive. If she had been inoculated, her beauty would be intact. She was determined to bring the Turkish technique home with her.

There was only one hesitation: She didn't trust English physicians to adopt the practice. Too many of them had made too much money for too long practicing old, ineffective ways of treating the disease. "I should not fail to write some of our doctors very particularly about it if I knew any one of them that I thought had virtue enough to destroy such a considerable branch of their revenue for the good of mankind," Lady Montagu wrote. "Perhaps if I live to return, I may, however, have courage to go to war with them."

After the Montagus returned to London, she got her war. When she began enthusing about the Turkish engrafting, the English medical community reacted with disdain. Their pushback was in part religious (what had these Mohammedans to teach a Christian nation?), part sexist (what could an untrained female teach a trained male physician?), and part medical. The common approach to dealing with smallpox in England in 1720 was based on the ancient system of balancing the four humors: blood, phlegm, black bile, and yellow bile. The theory was that when something threw these humors out of balance, disease resulted. Treatments were designed to bring them back to equilibrium. In smallpox, the pustules were obviously the body's attempt to balance itself by expelling vile matter from within. The physician's duty was to help nature do its work by subjecting patients to bleedings, laxatives, and induced vomiting.

Thus weakened, they died in droves.

The inoculation *à la Turca* enthusiastically described by Lady Montagu did not fit this framework. So they dismissed it.

IN THE SPRING of 1721, yet another smallpox epidemic began raging through London. This one was particularly deadly. Lady Montagu now had a daughter, born just before departing from Constantinople (and thus too small to get an inoculation at the time), and Mary was determined to protect this second child from the disease. The girl was now three years old, which might be just old enough for an inoculation. Lady Montagu called in Maitland, who had also returned home, to do it. The Scot again was reluctant; if anything went wrong it would be a major blow to his medical reputation. To shield him and to encourage others, witnesses were brought in to observe the procedure. Lady Montagu wanted this to be more than a private decision. She wanted her daughter's inoculation to be a public demonstration of its efficacy.

Since she'd been unable to have much effect on physicians, Lady Mary had taken to talking up the procedure with other members of her social set. She had friends in high places, up to the palace, including Caroline, Princess of Wales, wife to the heir to the British throne. Caroline made sure that one of the witnesses was the Royal Physician himself. The bewigged notables gathered and watched the procedure performed before their eyes, the nervous Maitland using his scalpel to cut small incisions in the girl's skin and depositing the pus from a mild case.

It went well, and Lady Mary's daughter breezed through the expected mild form of the disease, with her recovery observed by some of the leading medical men of the day. Mary encouraged people to call at her home to see her daughter, and they received a steady stream of visitors, some medical, some social. Soon, with the epidemic still raging, many of the aristocrats in Montagu's circle began asking to have their own children inoculated.

Foremost among them was the Princess of Wales herself. Caroline, the German-born wife of the future George II, was at the time mother to five small children, one of whom would one day inherit the throne. Caroline, like Mary, was also very intelligent. She corresponded with the great German thinker Gottfried Wilhelm Leibniz and other leading minds of her day. Voltaire called Caroline a philosopher in royal robes. It was no wonder she and Lady Mary hit it off. And after seeing what happened with Mary's daughter, Caroline was intent on having her own royal children inoculated.

Caroline of Ansbach by Enoch
Seeman, c. 1730

She began lobbying her father-in-law, King George I, for permission. And he refused. The king would not risk his bloodline to this foreign technique without further proof of safety. Caroline was forced to arrange a further experiment, this time on convict volunteers from Newgate Prison. In exchange for their help, the chosen prisoners would receive royal pardons.

Three male and three female prisoners duly underwent inoculation before an audience of a couple dozen scientists and physicians and were then kept under close watch. Within weeks, five of them developed the expected mild case of smallpox and recovered (the sixth, it turned out, had already had smallpox, so the inoculation did nothing). But did the inoculation truly make them resistant to the "wild" smallpox that was rampaging through London? To find out, one of the prisoners, a nineteen-year-old woman, was ordered to lie every night in the bed of a ten-year-old boy who was battling a severe case of smallpox. She cared for him for weeks and didn't come down with the disease. That was encouraging, but was it enough proof?

It was not. Another demonstration was arranged, this time using eleven London orphans as the test subjects. The results, again, were good.

The use of prisoners and orphans in these early experiments set the tone for medical experiments for the next two hundred years: When a new drug needed testing on groups of humans, it was easiest to go where the subjects had little power to object—where their actions and movements could be controlled, and where they could be observed over time. Prisoners and orphans were considered perfect; so, later, were mental patients and soldiers. Patients confined to hospitals were another possibility. It is only fairly recently, in historical terms, that physicians became concerned about things like informed consent.

In September 1721, the doors of Newgate were thrown open and six healthy, newly inoculated prisoners walked free. It was an historic moment. These tests on prisoners and orphans were the first "clinical trials," as we would call them today—tests of a new drug or medical procedure on groups of humans to see if they are safe and effective. Clinical trials are now a standard part of all modern drug testing. Every prescription drug today must be shown to be safe and effective in humans, and the only way to do that is to have humans take it. Clinical trials now commonly involve hundreds or thousands of patients, and the clinical trial industry is big business.

But in 1721, there were no such standards. All it took was a handful of doctors, six prisoners, and eleven orphans. Still, by the standards of the day, these were true scientific experiments. The tests were thought through ahead of time, run on multiple subjects, and carefully monitored, with observations recorded and results published. Others could then try the same methods and compare results. Medicine was turning into a science.

Mary's and Caroline's demonstrations had their effect. Inoculation caught the interest of more scientists and physicians, who slowly, tentatively, began to adopt the procedure.

But it took one more celebrity endorsement to get the public on board. It happened in the spring of 1722, when Princess Caroline finally received permission from the king to inoculate her two oldest daughters. This permission, notably, was granted only for the girls: Risking a potential male heir to the throne was too far for the king to go. The girls underwent inoculation and both survived. The public rejoiced.

This royal demonstration had two results. First, the nobility of England, in increasing numbers, arranged inoculations for their own children, setting off a ripple effect in which more and more physicians started providing it, and so it became available to more of the general public.

The second result was a countermovement, the start of a public reaction against inoculation—the direct ancestor of today's anti-vaccine activism.

The anti-inoculators of Georgian England made their arguments in pamphlets, newspapers, pubs, and coffeehouses. Some argued that the practice was foreign and barbarian; some were suspicious of its promotion (and in Turkey, even practice) by females; some considered it ungodly; and many deemed it dangerous. There was a political component, too: Because the royals were in favor of it, anti-royalists automatically viewed it with suspicion.

There was plenty of ammunition for the anti-inoculation forces. As the practice spread, a small fraction of the recipients went on to develop a more serious form of the disease. Some died. By 1729, according to one accounting, of 897 inoculations performed in England, 17 deaths resulted. This mortality rate, of around one in fifty, was far better than the one-in-four chance of dying from naturally transmitted smallpox, so a number of leading physicians continued to back the new procedure. But some of the public turned away from it, encouraged by clergymen who argued that only God had the power to determine life and death, and that inoculation was therefore un-Christian. By giving inoculations that sometimes killed, were not physicians acting as poisoners?

The anti-inoculation movement was fueled by vivid stories of failed procedures, patients who died, patients whose family members caught the disease from them and died, xenophobia, questions about criminality. Why were physicians being allowed to profit from this suffering?

Some doctors refused to do inoculations. Others tried to improve the procedure. The advent of inoculations marked a transitional period in medical history, when the two-thousand-year reign of one grand medical theory—the idea of the four humors—was giving way to new insights gained from the application of science. Doctors with one foot in each world tried to fit inoculation into the old structure. Pus formation was viewed as a good thing in the old system—"laudable pus" was a sign of healing—so English physicians preferred using scalpels in place of needle scratches for inoculations, making their incisions deeper, cutting through the skin and into the muscle, in part to ensure better pus production. Other holdovers from the old system included a continued emphasis on bleedings, purgings, and strict diets.

So arose the English variation on the Turkish procedure. No longer was inoculation a quick scratch followed by a period of isolation while the mild disease rose and faded. English physicians insisted on lengthy, complex regimens of preparation, giving children days or weeks of laxatives, bloodletting, and special diets before inoculation. This made the procedure more difficult, more time-consuming—and more profitable for physicians. Because most early adopters of inoculation were well-heeled members of the aristocracy, they could afford to spend a great deal. Prices were inflated accordingly.

One of the children who underwent the procedure was an eight-year-old orphan who wrote later about being "prepared" for weeks, bled and purged repeatedly, put on a low-vegetable diet, and confined to an "inoculation stable" with other boys. He was so weakened by the time of his exposure to the pox that he fell severely ill and was kept in the stable for weeks before he was finally released. It was a horror show that stayed with the boy for the rest of his life. His name was Edward Jenner.

But by Jenner's day—the latter half of the 1700s—most physicians at least accepted that inoculation was the best tool they had to fight smallpox. And they were getting better at it, gradually giving up the deep incisions and bleedings, moving back toward the Turkish method. The easier and cheaper inoculation became, the more it was used. There was talk of government support for public inoculation.

The practice spread to America and throughout Europe. In America, a black slave who had been inoculated by his tribe in Africa helped convince his master, Cotton Mather, to push for its use. In Russia, Catherine the Great was secretly inoculated in 1768 by a physician (who was so nervous about a possible failure that he kept horses at the ready in case he needed to escape). Thousands of people were undergoing the procedure.

Lady Mary had won. She went on to live a long and notable life, hobnobbing with the great minds of her day (she was so admired by the great poet and essayist Alexander Pope that she reputedly had to spurn his advances), falling in love with a brilliant Venetian count (for whom she left her husband), traveling through Europe, and continuing to earn fame for her writing. Her son, the boy she had inoculated in Constantinople, lived a disappointing life, becoming a wastrel and a gambler. Mary's daughter, the girl who was made

L.L. Boilly, Vaccination, 1807 by Louis Boilly. Wellcome Collection

into a medical demonstration, married a future prime minister of England.

Lady Mary Wortley Montagu should have been lauded after her death in 1762 as a pioneer of medicine. But her great achievement, the introduction of inoculation to Europe, remained little-known until recently. The world's attention and its honors went instead to Edward Jenner, that boy who had suffered so terribly in the inoculation stable, and who would go on to become famous as the father of vaccination.

DAIRYMAIDS HAD THE best complexions. This was something people in the countryside knew: English dairymaids, the girls who milked the cows every morning, tended to be rosy-cheeked, creamy-skinned, and—this was the important part—unscarred by the pox. Maybe it was their diet, richer than most in milk, cream, and butter. Or maybe it was something else. Cows' udders were sometimes spotted with a mild disease called cowpox. It looked a bit like smallpox but was no real threat. Dairymaids often picked it up on their hands while milking, getting a spotty rash that passed after a few days. After that they rarely got smallpox. Dairymaids, therefore, were commonly used as nurses if someone on the farm came down with the disease. These things were well known to country folk.

Farmers could get cowpox, too. It happened in the mid-1700s to a tenant farmer near Dorset named Benjamin Jesty. He was young at the time, and like a lot of farm folk around Yetminster village he suffered the rash, it went away, and he didn't think much about it. Jesty went on to become a pillar of his community, a farmer known for his hard work, common sense, and growing prosperity.

Among his many friends and acquaintances was John Fewster, an area physician who practiced inoculation. Fewster knew about the local belief in the dairymaid-cowpox-smallpox link. He had once given a little talk in London on cowpox's seeming ability to prevent the more serious disease. It hadn't gotten much attention.

Fewster might have theorized, but it was farmer Jesty who put the idea into action. In 1774, as a smallpox epidemic threatened his area, he didn't worry about himself—he'd had the cowpox. But his wife and two young boys hadn't gotten that mild disease, nor smallpox. The nearing epidemic could kill them. So Jesty determined to give them the same protection he had gotten. He asked around and learned of a cow in a neighboring dairy herd that had a case of cowpox. He gathered up his family and led them in a trudge across the fields to the infected animal. There he scraped and poked some of the cowpox material from the animal's udder and, using a darning needle, scratched it into the arms of his wife and children.

This animal-to-human transfer did not go well at first. His wife's arm got infected, and a doctor had to be called in to treat her. His neighbors found out and, hooting and jeering, pelted him with mud and stones for his affront to God.

But it worked. His family, all three of them, got mild cases of cowpox. And later, when the smallpox outbreak fell on his village, it did not touch them. Jesty had likely saved their lives. But he was a humble man who wanted to keep good relations with his neighbors. So he didn't brag about it. He went back to farming.

The story came out only later, when Jesty was celebrated as the first person to perform what would come to be called a "vaccination" (from *vacca*, Latin for "cow").

This was a term invented some years after Jesty's experiment, by the man who would take the lion's share of credit for its discovery: Edward Jenner. In the 1790s, decades after Jesty's trudge across the fields, Jenner did the careful scientific work needed to convince the world that vaccination with cowpox was both significantly safer and more effective than the old method of inoculation with smallpox; it was Jenner who, after a period in which his ideas were first attacked, then accepted, earned worldwide fame. As the scientist Francis Galton later put it, "In science the credit goes to the man who convinces the world, not to the man to whom the idea first occurs."

And Lady Montagu's pioneering efforts—like the efforts of many other women in the history of science—were for the most part ignored.

IN 1863, just hours after giving the Gettysburg Address, Abraham Lincoln fell ill with what most historians think was smallpox. He recovered after a four-week illness; his personal valet, however, died from the disease.

Despite all that Mary Montagu, Benjamin Jesty, Edward Jenner, and others had taught the world about preventing smallpox, the disease still ravaged much of the world, and it would continue to do so through the next one hundred years. During the twentieth century alone, smallpox is estimated to have killed three hundred million people worldwide—more than twice as many victims as all of the century's wars and natural disasters combined.

But smallpox vaccination was having an effect. The more people were vaccinated, the fewer victims there were to spread the disease. The nations that vaccinated most aggressively, mandating it for schoolchildren, were able to cut down the number of cases all the way to zero. The last case of wild smallpox in the United States was in 1949; in North America, 1952; in Europe, 1953. It was clear that if the same aggressive vaccination efforts were undertaken in every nation, there was a good chance the disease could be cleared from the planet.

It turned out that smallpox, the greatest of all killers, was also the perfect candidate for eradication. For one thing, it was easy to track. The symptoms were obvious after a couple of days, so patients could be identified and isolated before it spread very far. Also important was the fact that the strains that infect humans don't infect other animals. There was little or no possibility of an "animal reservoir" of smallpox hiding in some remote spot, waiting to reinfect humans—which can happen with other diseases (like yellow fever, which can also infect monkeys, and then jump back to humans). Finally, recent smallpox vaccines—much more effective than Jenner's cowpox inoculation—are very effective, easy to use, and safe, making it easy to protect large populations in a short time.

Today we know much more about how vaccines protect us. Montagu, Jesty, and Jenner had made their discoveries through simple observation: They saw what worked, and they tried to make it work better, for more people. They didn't know *why* it worked, because they didn't know what caused smallpox—or any other contagious disease.

Those discoveries waited until the last half of the 1800s, when Louis Pasteur, Robert Koch, and others showed that many diseases were caused and spread not by a derangement of the humors, but by invisible living organisms called "germs." Germ theory hit medicine like a bombshell, exploding old theories and clearing the way for new approaches to healing. Among them were more vaccines for other diseases like rabies, anthrax, measles, and eventually polio. The right vaccine could work miracles for certain diseases.

But not all of them. A number of vaccines were tried that simply didn't work well. It depended on the specific disease. From the 1880s through the 1930s, scientists tried to figure out why. Why did some vaccines work and others didn't? Why did vaccines work at all?

The answer was found within the body's own defense mechanisms. Hand in hand with germ theory and vaccine development, we began to learn more about the body's immune system, the elaborate, finely balanced, many-player system that allows our bodies to identify, target, and destroy invading organisms like bacteria and viruses. It turned out that Lady Mary's inoculation and Jenner's vaccine acted like wake-up calls for the immune system by delivering small doses of a virus (an infective organism even smaller than bacteria; the first virus was identified in 1892). Once the invader was identified, the body was able to remember it and mount a very fast defense if it showed up again. It was immune to the disease.

Smallpox, it turned out, was caused by two strains of the variola virus, one very dangerous (*Variola major*) and one milder (*Variola minor*). Vaccines work very well against both—better, in fact, than most vaccines work against other diseases. Each contagious disease is different. Flu viruses, for instance, come in many strains that mutate and change every year, so vaccines can be less effective. Malaria is caused by a much different pathogen, a parasite. There is no highly effective vaccine for malaria. Some viruses and germs, like the virus that causes AIDS, have learned how to hide from the immune system, making vaccines less effective. And so forth.

But smallpox vaccines worked so well that by the 1960s, global health initiatives were on the verge of wiping out the disease. It was a tremendous effort. Workers trekked through jungles and were airlifted to mountain villages, vaccinating everyone they could get to in increasingly remote areas of Asia, South America, and Africa. Their goal was something new to medicine: not just controlling a disease, but ridding the world of it forever.

And it didn't take long. In 1977 Ali Maow Maalin, a twenty-three-year-old Somali health worker and hospital cook, earned a place in history as the last person on planet earth to be infected with naturally occurring smallpox. Somalia, with its nomadic tribes and remote terrain, was one of smallpox's final refuges. When Maalin came down with the disease he was immediately put under quarantine; everyone who had come into contact with him was checked for recent vaccination and carefully monitored. He lived through his disease and went on to devote his life to fighting polio. World health experts took a deep breath and watched. For months—well past the time most researchers thought the virus could live without a human host—there were no more cases anywhere.

Victory was declared. Smallpox, the deadliest disease of all time, was gone.

OR SO PEOPLE thought.

In 1978 Janet Parker, a middle-aged photographer in Birmingham, England, came down with what she thought was a cold. Then a rash arose. And turned into pustules.

Her doctors were astonished. No one had seen a case of smallpox in Britain for decades. But the signs were unmistakable. Then they learned that she had been working in a local hospital, where her job was taking pictures of tissues and organs for doctors' files. She developed her film in a darkroom just above a laboratory where a medical researcher named Henry Bedson was doing studies—on smallpox.

The virus was gone from the natural world, but a few samples had lingered on, frozen and locked away for posterity (and scientific study) in a handful of laboratories scattered around the world. Bedson's was one of them.

When the story emerged later, it was found that Bedson's smallpox lab was in trouble: Authorities had warned him that his facility did not meet international standards for safety and were threatening to shut him down within a few months. At the time Parker picked up the disease, Bedson had been rushing to get results while he could.

No one knows how it happened, exactly. The virus might have gotten into the hospital air ducts, or maybe it was passed along on contaminated clothing or equipment—even an official inquiry later couldn't pin down the route—but somehow Bedson's virus got into Janet Parker.

This was a medical disaster in the making. Her home was sealed off and sterilized. Her vaccination records were checked: She had gotten a smallpox vaccination, but it had been twelve years earlier. To keep up immunity, smallpox vaccinations should be renewed every few years. But since there was no more smallpox around, she—like many others—neglected to stay current. There had not been a case of smallpox in the UK for so long that people weren't bothering to get vaccinated; many young people had no immunity at all.

Parker was quickly quarantined, along with everyone that health authorities could find who had been in contact with her, around five hundred people in all, including Parker's parents and the ambulance driver who drove her to the hospital.

Suddenly British medical care was thrown back seven decades. Where could all of Parker's contacts be quarantined? There was an old "fever hospital" built in 1907 to isolate the most serious cases of infectious disease—a place so little used in the 1970s that its staffing was down to just two employees. It was scrubbed, refurbished, and quickly brought back to life. Many of Parker's contacts were housed there and monitored for signs of disease.

Most of the attention went to Parker herself. Her condition worsened. The pox was everywhere on her body, from her scalp to the palms of her hands and the soles of her feet. Her breathing became labored. The scene began to turn nightmarish: Parker's mother came down with it, too. Her father, isolated in the same hospital, worrying about both his daughter and his wife, had a heart attack while visiting Parker's room. He died within days.

In the middle of it all, Henry Bedson, the smallpox researcher, walked into the garden shed of his house and cut his own throat. His suicide note read, "I am sorry to have misplaced the trust which so many of my friends and colleagues have placed in me and my work and above all to have dragged into disrepute my wife and beloved children. I realize this act is the last sensible thing I have done but it may allow them to get some peace."

Ten days later, smallpox killed Janet Parker.

Her body was treated as a biohazard. Her funeral was overseen by health authorities, the cortege accompanied by police in unmarked cars. Mourners were kept away from the body, which was burned in a specially monitored crematorium. The crematorium was scoured afterward by medical technicians.

There were official inquiries, a debate in Parliament, and finally action by the World Health Organization. Smallpox, it was decided, was clearly too dangerous to be researched in so many labs. If it got out of containment, the risks were too high. Within a few years of Parker's death, virtually all lab stocks of smallpox virus in the world were destroyed. The only samples of the Speckled Monster remaining today are kept in two tightly locked-down laboratories, one at the U.S. Centers for Disease Control in Atlanta, one at the State Research Center of Virology and Biotechnology in Koltsovo, Russia.

At least as far as we know. There is no guarantee that there might not be rogue stocks of the virus held secretly in other places. The fracturing of the Soviet Union in the 1990s led to concerns about the security of the smallpox samples stored there; the rising threat of international terrorism since 2001 has increased the worry. In 1994 a research team published the complete genome of the smallpox virus, and with the fast-advancing tools available to manipulate genes, nothing says that a rogue lab might not someday reconstruct the living virus.

No one has had smallpox for forty years. No one has treated it, and only a small fraction of humans are immune to it. In the United States, routine smallpox vaccinations for every child were discontinued in 1971, and today, smallpox vaccination is required only for U.S. military members posted in Korea and a few other specialized cases. Right now, we are about as susceptible to the pox as an Aztec or Inca, or a British toddler in 1700.

To counter the threat, the United States started a crash program after 9/11 to make and stockpile millions of doses of the smallpox vaccine—enough to quickly vaccinate everyone in the nation if necessary.

It all comes down to risks and benefits. Vaccination risks are low and complications are rare, but they do exist. Because the risk of smallpox today is close to zero, running even the small risk of side effects from bringing back routine vaccination seems unwarranted. But we keep vaccine on hand just in case.

That same risk/benefit analysis should be applied to all vaccinations. Some decisions, like whether to have a flu shot, are left up to individuals. The flu is generally mild, and vaccines to prevent it are nowhere near 100 percent effective, so whether to get a flu shot or not is up to you. Same for conditions like shingles and herpes virus. Vaccines for these diseases are available, safe, and a good idea for high-risk populations—but it's your choice.

Things change when the disease is more dangerous. Health experts make sure that vaccines for severe diseases like diphtheria and tetanus are mandatory for children. Here the major benefits of avoiding the disease far outweigh the minor risks from vaccination; it's clearly in the interests of public health to mandate vaccination.

That doesn't mean that anti-vaccination activism has gone away. If anything, it's stronger than it's been in the past century, fueled by rumors and scares exploding across the Internet. In part, today's anti-vaccination movement is rooted in vaccination's success. The diseases we vaccinate against now seem like harmless ghosts for the most part, robbed of their power to terrify—because vaccines have made them a thing of the past. Few people living today have ever seen a case of smallpox, or diphtheria, or polio. They've never lost a brother like Lady Mary did, or a daughter like Janet Parker's mother, to these killers. Our sense of risk is lessened to the point where, to many, the benefits of vaccination seem small—so small that even vague risks from vaccines seem large.

That is, in my mind, dangerously wrongheaded. The more people decide not to vaccinate, the larger the pool of people without immunity grows—and the faster a resurgent disease can spread. The reason that smallpox has been cleared from our planet is that when enough people are vaccinated, and in the absence of another animal host, the virus has no place to multiply—no way to spread. It dies out. If enough people are vaccinated, the danger approaches zero. That's the benefit of what's called "herd immunity."

The triumph over smallpox was hard-won. Immeasurable suffering has been prevented. Hundreds of millions of deaths have been averted. Today, other killers like polio are within reach of eradication. Lady Mary, with her independence, her wit and influence, and her perseverance, helped open the door to these miracles. We should honor her common sense, her bravery, and her memory by continuing her work.

CHAPTER 3

THE MICKEY FINN

OPIUM AND MORPHINE were natural products, made from a plant. So was almost every other drug available to physicians in the mid-nineteenth century (along with a few non-plant substances like mercury). They were all refined from nature.

But that was about to change. Science in its modern sense—that is, based entirely on observation, experimentation, publication, and replication—was just beginning to make a mark on the world of drugs. The old structures built to explain health and the natural world—a tangle of ancient theories from Rome and Greece, leavened with some Arabic insights, and shoehorned into a Christian framework—had already given way. Now the new sciences were about to release a flood of new drugs.

No scientific discipline in the mid-nineteenth century was as dynamic, as revolutionary, or as important to medicine as chemistry. At a very simple level, chemistry is about how atoms join together to make molecules, and how those molecules behave with one another. It was here, at the level of molecules, that chemists in the 1800s crashed headlong into religion.

It had to do with the definition of life. In the West, the dividing line between life and death had long been set by Christianity. The difference between the two was the presence of a holy force, a God-given spark that separated dead rocks from living creatures. This was not only a religious idea; many scientists around 1800, for instance, believed that the chemicals found in living things—organic chemicals—were fundamentally different from other chemicals. There was some good evidence to back them up: While chemical reactions in a lab, for instance, could in most cases be reversed, with reactants changing to products and products changing back to reactants, reactions using chemicals made in living bodies, it was thought at the time, could not. You could not turn wine back into grape juice or unfry an egg. The organic chemicals involved in the processes of life, it was thought, must have something in them that was different from other chemicals. Their actions could not be treated or studied the same way, and thus were lumped together as the new field of organic chemistry. There was something unique about them; they operated by a different set of rules, were touched by something else—perhaps that vital spark.

This idea of vitalism permeated chemistry in the 1700s and early 1800s. Chemists took sides: Some believed that all chemicals were the same, and that eventually organic chemicals would be seen to conform to the same rules that governed the rest of chemistry. There was no vital spark, no mystical something that separated life from death. Others argued that there was certainly something different, more special, perhaps divine about chemicals involved in living organisms.

Most healers of the day continued to believe that life was permeated with a special spirit, and that a balance and flow of vital forces in the body guided health. These "special forces" ideas ruled Western medicine for centuries under the general heading of the four humors, while in China it was seen as the flow of chi. Today, it lives in the alternative healer's belief in subtle energies.

But not in chemistry. The idea of this hard-and-fast dividing line between the quick and the dead took a literary blow in 1818 with the publication of Mary Shelley's novel *Frankenstein; or, The Modern Prometheus*, with its doctor-protagonist playing God by restoring life to dead tissue, and then a more scientifically important punch in 1832, when the German chemist Friedrich Wöhler showed that he could synthesize one of the substances thought to be made only in living bodies, the molecule urea, entirely in his laboratory from the combination of two dead chemicals. This seems like a small thing now. But at the time it was a big deal. Science, with its ever-more-powerful array of facts and techniques, was blurring the line between life and death. Scientists were crossing a threshold.

Wöhler's great friend—and arguably even greater chemist— Justus von Liebig took the next steps. Liebig was a phenomenon of science, a true genius, a great teacher, who was passionate about

Justus von Liebig. Photograph by
F. Hanfstaengl. Wellcome Collection

applying chemistry to everything—especially living processes. This German chemist was fascinated with the ways in which living organisms interacted with the nonliving world, especially the chemistry of that interaction. He was the first to show, for instance, that growing plants required certain mineral elements—nitrogen, phosphorus, potassium, and so forth—in order to thrive. In other words, he figured out how fertilizers work. He was the father of agricultural chemistry. And this difficult, demanding, opinionated man also had a lifelong interest in drugs. He became doubly famous as the father of clinical chemistry, the use of chemistry in medicine.

In fact, what Liebig was doing was demonstrating that nutrition, growth, the processes of life itself did not come from God alone, but from chemical changes. He summarized his ideas in his 1842 book, *Animal Chemistry*.

After Liebig, most scientists considered that living processes could effectively be reduced to a series of chemical reactions. The body could be picked apart in finer and finer detail, reduced all the way down to the level of molecules. This reductionist approach has guided much of the study of life ever since. God no longer sets the terms of the argument.

Along the way, Liebig made a lot of interesting new chemicals. One of them, chloral hydrate, first emerged from his lab in 1832. This entirely synthetic chemical could not be found in the body; it had never existed on earth, as far as anyone knew, until Liebig made it. And yet it was destined to be used as a medicine.

Liebig didn't know that. He never thought of using it as a medicine. He was tinkering, playing with molecules, learning what transformed one into another. He found, for instance, that he could turn chloral hydrate into a heavy, sweet-smelling liquid called chloroform, the fumes from which could knock a person unconscious. By the 1850s, chloroform was being tested as a way to put patients to sleep before surgery. But it was too hard to handle, too dangerous—it was easy for patients to breathe in too much, and accidental deaths happened on the operating table—so researchers set it aside and started looking for alternatives. Liebig had shown that he could transform chloral hydrate into chloroform in his lab, so might the same thing happen in the body? Could chloral hydrate be a safer alternative to chloroform? They started testing it on animals.

Chloral hydrate is a solid at room temperature, but it can be made into a more easily administered liquid by simply mixing it with alcohol. In either form, solid or liquid, it was found in the 1860s to be great at putting humans to sleep. It had been around too long to be patented—its medicinal use didn't start until decades after Liebig first made it—but it was made by a number of firms and widely used.

Although natural drugs like opium could make users sleepy, they also had other effects. This made chloral hydrate, in the eyes of many historians, the first true sleeping pill, a class of drugs physicians called "hypnotics." A little chloral could calm patients down, a bit more could help ease them to sleep, a lot could knock them out. By 1869 it was being sold as a sleeping aid and a way to soothe patients before surgery. More than being the first hypnotic, chloral was the first widely used, totally synthetic drug.

Within a few years it grew into an international fad. Like morphine, it was taken both as a medicine and as a recreational drug. Nervous Victorians used it to tranquilize themselves. Insomniacs devoured it before bedtime. Partygoers played with its effects. As the New York Times reported from London in 1874: "Chloral is the now-fashionable hypnotic, the means by which balmy sleep, nature's sweet restorer, is wooed."

It was also dangerous. As use spread, so did reports of accidental overdoses and use in suicide. And worse.

IN THE FALL of 1900, a seventeen-year-old girl named Jennie Bosschieter took an evening walk from her family's working-class apartment in Paterson, New Jersey, to get baby powder for her niece. She never came home. The next morning a milkman found her body on the shores of the Passaic River. She'd been raped. And poisoned. An autopsy showed that she'd overdosed on chloral hydrate.

The story that emerged became a Gilded Age cause célèbre. A few days after Bosschieter's body was found, the driver of a horse-drawn hack admitted that he'd picked her up at a saloon the night before, when she was carried into his carriage by four men who brought her out of the saloon's side door. She was unconscious but alive. The men directed the driver to an isolated spot out in the country, where, he told police, they spread a blanket and repeatedly assaulted the girl. The only time they paused was when she vomited. When they got her back into the carriage, she was limp and unresponsive. Her attackers got worried. These four young men seemed well-connected; they directed the carriage driver to the home of a leading local physician who was a family friend of one of the girl's attackers. But it was too late. The girl was dead. They carried her body back to the carriage, ordered the driver to the river, dumped the body, and gave the driver $10 to keep his mouth shut.

It wasn't enough. A few days later the driver went to the police, the police went to the doctor, and the doctor gave up the young men. They were all from respectable, wealthy local families. One of them was the brother of a judge.

The four men blamed the victim, saying that she'd joined them voluntarily, flirted with them, was drunk and throwing her arms around them. They bought her absinthe and champagne, they testified, but knew nothing about any chloral. They simply took her for a carriage drive, got worried when she passed out, then panicked when she died. They couldn't explain why her underwear was missing. Or how a bottle with chloral in it was found near the body.

The better elements in town chose to believe the young men, and rumors began flying about this loose factory girl, a working-class teenage strumpet who had cast a spell over their favorite sons. A socialist newspaper came to Bosschieter's defense, framing her death as an attack on the working class by upper-class degenerates. The newspapers loved it.

The eventual trial was a public spectacle, the courtroom packed and buzzing. Hundreds who'd been turned away milled about outside, yelling at witnesses as they arrived.

Under cross-examination the four young men, counseled by some of the area's best attorneys, stuck to their story. But the evidence was too damning. After three days, they were all convicted of second-degree murder. Three of them were given thirty-year sentences. The fourth finally confessed to the crime, provided details, and was given fifteen years. All of them were released after serving little more than half their sentences, thanks to years of what one newspaper called "relentless pleas for clemency on their behalf by Paterson's influential class."

JENNIE BOSSCHIETER had died from a mixture of chloral hydrate and alcohol commonly called "knockout drops." It was the original date-rape drug. And it found other uses as well.

There was Mickey Finn, for instance. Now more a phrase than a figure, Finn was likely a real person, the bartender and manager of a saloon that operated around the turn of the century on the South Side of Chicago. In 1903, a prostitute named "Gold Tooth" Mary Thornton testified that one Michael Finn, manager of the Lone Star Saloon, was poisoning his customers and robbing them. The simple system worked like this: Finn or one of his workers, a waiter or a "house girl," would slip chloral hydrate into a likely customer's drink; when the drug took effect, the semiconscious customer was escorted (or carried) into a back room, robbed, and dumped in an alley. Later the victim couldn't remember much.

Finn was caught and his bar shut down, but the idea of "slipping somebody a Mickey" was just getting started. Knockout drops would become part of the criminal fabric of America.

Chloral's legitimate uses, most of them in mental hospitals, were even more important. Sometimes mental patients were out of control, manic, thrashing—a danger to themselves and those around them. In the old days, attendants had used force and restraints like straitjackets to manage them, and opium, morphine, even cannabis to calm them. But chloral was better, faster, less liable to cause hallucinations, and a more controllable way to knock patients out. In smaller doses it could calm agitated patients and ensure a quiet night of sleep for patients and attendants alike. It's no wonder that for three decades around the turn of the century, you could tell you were in a mental hospital even if you were blindfolded. It was the smell—the pear-like smell of chloral from the breath of patients. The mental wards reeked of it.

The chloral era lasted until around 1905, when chemists came up with even better synthetic drugs, barbiturates, followed in the 1950s and 1960s by early forms of today's tranquilizers, and the more powerful antipsychotics (see the chapter on chlorpromazine, page 123).

We now have hundreds of types of improved sleeping pills, better relaxants, and more varied drugs for criminals to lace into their victims' drinks. Chloral still gets prescribed and used (it was, among other things, in the cocktails of drugs that killed Marilyn Monroe and Anna Nicole Smith), although it's now a minor player.

But it earned its place in history. Chloral, the first widely used, totally synthetic drug, broke new ground. It proved that scientists working with test tubes in laboratories could make medicines that could match or exceed the power of those made by nature. Its eager adoption by mental health experts, its enthusiastic use by sleepless members of the public, even the subsequent press attention to its lurid criminal use all pointed toward profits that could be made by exploring other laboratory-made drugs.

Liebig's and Wöhler's scientific heirs, the generations of organic chemists who came of age in the late 1800s and early 1900s, became masters at tinkering with molecules that had an effect on the body, adding a few atoms here, taking away a few there, tailoring them for specific purposes. The more new chemicals they made and tested on animals and humans, the more they learned about what worked to promote health and what didn't. Along with the blooming of chemical industries in general, a few chemists began devoting themselves to finding new synthetic drugs.

Knockout drops helped give birth to the behemoth we now call Big Pharma.

CHAPTER 4

HOW TO SOOTHE YOUR COUGH WITH HEROIN

THANKS MOSTLY TO the joys of shooting morphine, the United States in 1900 was estimated to have around 300,000 opiate addicts out of a total population of about seventy-six million, or about four addicts per thousand people. That means that in rough terms the rate of opiate addiction in America in 1900 was about the same as it was almost a century later, in the 1990s. In the past twenty years, of course, the rate of opioid addiction has shot up considerably. But many things were similar about the epidemic then and now. Then, as now, overdoses were killing thousands every year. Then, as now, everyone knew about the dark side of opium-derived drugs; everyone was reading news reports about the suicides and overdoses, the addiction and despair. And then, as now, no one quite knew what to do.

The main difference was that in 1900 opium- and morphine-laced drugs were available without a prescription. You could buy a dose of morphine at the corner drugstore.

But in the face of an addiction epidemic, a growing number of physicians, lawmakers, and social activists demanded that something be done to control the drugs. Total prohibition was not an option. Morphine was too valuable a medicine to ban entirely. But pressure grew for some kind of regulation.

While politicians argued about the legalities, scientists searched for something that would make the legalities meaningless. They wanted to find some new form of morphine that would have all of its painkilling power with none of its addictive risks. This magic medicine became the holy grail for drug researchers. Chemists began to study and change the morphine molecule, adding a side chain here, taking an atom or two away there, continuing the quest.

Every year chemists were getting better at what they did. The decades around 1900 were a golden age for chemistry, especially the subfield organic chemistry, the science of carbon-containing molecules like proteins, sugars, and fats—the molecules of life. These wizardly chemists seemed able to make almost any variation they wanted of almost any molecule in the body. They were learning how sugars are built, how foods are digested, how enzymes (the catalysts of biochemical reactions) work. They could shape molecules the way other people shape wood or metal. They could, it seemed, do anything.

But morphine resisted them. A typical failure happened in London in 1874, when a chemist tried adding a little side chain of atoms (an acetyl group) to morphine. This British researcher was one of many looking for that magic combination, and he thought he might be on to something promising. But when he tested his new chemical on animals, he came up with nothing.

Animal testing is an imperfect art. Lab rats, dogs, mice, guinea pigs, and rabbits have different metabolic systems than one another and that of humans, and so can react differently to new drugs. Plus—and this is very important—they can't tell researchers how they're feeling. Without knowing that, scientists have to come up with other ways of testing the animal's reactions, trying to gauge the effects of drugs. Sometimes that's easy—like seeing if an infection has cleared up. Sometimes it's hard—like trying to measure the depth of depression in a rat.

Still, testing in animals remains one of the best ways researchers have to see if a new drug is poisonous, and to get at least a rough idea of its effects.

And so the London chemist in the 1870s gave his new acetylated morphine to animals. And nothing happened. It wasn't poisonous when given in low amounts, but neither did it seem to be doing anything. It was a dead end, like most experiments. He wrote a short journal article about his results and went on to other things.

There it sat for two decades, during which platoons of other chemists continued working with morphine and the other major alkaloids—opium, codeine, and thebaine—taking them apart and piecing them back together with new atoms, creating hundreds of variations. And the grail failed to appear. The greatest organic chemists in the world, with all their advanced techniques, were getting nowhere.

THAT IS, UNTIL just before the turn of the century. In the late 1890s, a dye-making firm in Germany decided to branch out. The Bayer company already had a stable of chemists whose job it was to turn coal tar (a waste product from making the gas that lit the Gaslight Era) into valuable chemicals like synthetic dyes. After Queen Victoria wore a mauve dress in 1862—a new shade made in a chemist's lab—synthetic fabric dyes became a fad. Chemists started making a dazzling rainbow of new colors out of coal tar. Everybody in the dye game made money. But by the 1890s there were a lot of dye makers in Germany. The market was getting crowded.

So Bayer turned its chemists to the task of exploring another moneymaking line of chemical products: drugs. Inspired by the success of synthetic drugs like chloral hydrate (see page 78), Bayer was determined to find more laboratory chemicals that could treat more diseases. The decision to move into drug-making was somewhat risky, but the rewards were potentially huge. The basic approach was the same for dyes and drugs: start with a common, relatively cheap natural substance (like coal for dyes, or opium for drugs), then allow organic chemists to alter the molecules in it until they turned them into something much more valuable. These newly created chemicals could then be patented and sold at a huge markup.

Soon after Bayer made the move into drugs, one of the company's young chemists, Felix Hoffmann, struck gold twice. In the summer of 1897, he, too, started attaching acetyl groups to molecules. When he did it to a substance isolated from willow bark (the bark had long been given as an herbal medicine to patients with fevers), he created a new and effective fever reducer and mild painkiller that his company named Bayer Aspirin. And when he linked the same acetyl side chain to morphine, just like the London chemist had done decades earlier, he came up with the exact same molecule that the British had already tested and discarded. But Bayer stuck with it, testing Hoffmann's acetylated morphine on more kinds of animals and interpreting the results more positively. They even rounded up a few young volunteers from the factory to test the drug on humans.

And the results were amazing. The German workers reported feeling really good after they took Hoffmann's new drug. No, better than good, great: They felt happy, resolute, confident, heroic.

That was enough for Bayer to give out some of the experimental drug to two Berlin physicians, with instructions to try it with any patients they thought appropriate. The results were, again, impressive. Bayer's acetylated morphine could ease pain, like morphine, and also turned out to be great at quieting coughs and managing

sore throats. Tuberculosis patients given the new drug stopped hack-
ing up blood. It had the pleasant side effect of raising spirits and
reviving a sense of hope. No serious complications or side effects
were noted.

That was all Bayer needed to hear. Enthused, the company
made plans to put their new wonder drug on the market. But first
they had to come up with a catchy trade name. The company considered calling it Wünderlich, the wonder drug. But in the end they
decided on a riff of the German word *heroisch*, or "heroic." Their new
drug would be called Bayer Heroin.

Their tests showed that it was up to five times stronger than
morphine and far less habit-forming, ten times more effective than
codeine and far less toxic. It looked to Bayer experts like Heroin had

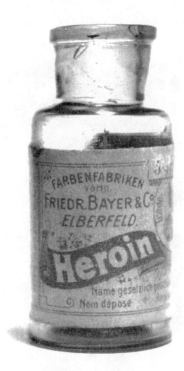

Bayer heroin, c. 1900

an additional, unusual ability to open up airways in the body, so they started selling it primarily for coughs and breathing disorders, and secondarily as a cure for morphine addiction. Patients happily gave up their morphine for Heroin. They loved the new drug. So did doctors. Use spread. For $1.50 users around the turn of the century could put in an order from a Sears-Roebuck catalog and receive back a syringe, two needles, and two vials of Bayer Heroin, all in a handsome carrying case. Early scientific presentations touting the success of Bayer Heroin spurred standing ovations.

But there was a problem. Because Bayer hadn't discovered Heroin—the molecule originally had been made by that London chemist two decades earlier—the drug's patent protection was weak, and other drug companies soon started making it. It lost the capital *H* used by Bayer and entered the wider world of drug-making and drug-hawking. Heroin-laced cough lozenges sold by the millions.

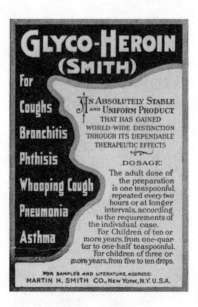

Advertisement from 1914 for cough medicine containing heroin

Elixirs containing heroin were said to be safe for all ages, even infants. The drug was added to one over-the-counter cure after another, touted as a treatment for everything from diabetes and high blood pressure to hiccups and nymphomania (the nymphomania application, at least, had some basis in reality: Heroin, as any addict can tell you, drains the sex drive). In 1906, the American Medical Association approved heroin for general use, especially as a substitute for morphine.

Without the ability to patent its new miracle drug, Bayer soon moved away from heroin, and the company stopped making it entirely around 1910. But by then Bayer Aspirin's massive global success was bringing in so much money that the company redoubled its focus on drugs. Dyes were put on a back burner, and pharmaceuticals moved to the front.

As heroin spread, physicians quickly figured out a couple of not-so-great things about the new drug. The first was that Bayer's idea about it being good for the respiratory system was wrong—the drug did nothing special to open up airways. The second was that heroin was no answer for morphine addiction, any more than morphine had been for opium addiction. Instead, the new drug was found to be very, very addictive. It was the story of morphine all over again: Physicians began seeing more heroin addicts in their offices, and newspapers began running more reports of overdoses. Heroin differed in some ways from morphine, but not in the important ways. Each refinement of opium, each new version, seemed only to increase strength without reducing addiction. Opium and all its children—morphine, heroin, and today's newer synthetic opioids alike—are all bewitching drugs, very good at easing pain, good at making users feel great (at least at first), easy to start, and, after a period of habituation, extremely hard to stop.

The term "drug addict" first began showing up in medical texts around 1900, at the same time the phrase "drug fiend" began seeing wider use in newspapers. (And another note about terms: "opiates" are drugs derived directly from opium, like morphine and heroin, while "opioids" is a broader term that includes today's synthetic painkillers as well.)

The problem went beyond opiates. There was also legal cocaine (used widely in hospitals and dentists' offices and, very briefly, as a minor ingredient in Coca-Cola); legal cannabis (a not uncommon ingredient in patent medicines); and legal anesthetics like ether and nitrous oxide (laughing gas). There was chloral hydrate and Bayer's popular new barbiturate sleeping pills, both for sleeping. Every year a new raft of drugs showed up, with extravagant claims and little regulation.

In the years just before World War I, America woke up to the fact that it had a drug problem. Muckraking journalists began exposing the dangers of drugs, from patent medicines to chemical-laced cosmetics. Drugs were tearing families apart, spurring addicted women to prostitution and addicted men to robbery, bringing down financial ruin and personal disgrace. The antidrug movement gathered up medical experts and ministers, housewives and newspaper editors, do-gooder politicians and hard-nosed police alike to form a broader social movement for drug control. Part of it grew out of the Bible-fueled temperance movement against alcohol. Part of it was rooted in the reform-minded Progressive politics of the day. A mixture of moralism and medicine with a dash of racism—*Look at those Chinese opium dens, marijuana-dazed Mexicans, and drug-crazed Negroes*—propelled the antidrug campaigns.

It all came to a head during the years when Theodore Roosevelt was president. TR was a Progressive, dedicated to clean government and decisive action. He, like many people, felt that patent medicine makers were bilking the public with overblown claims for secret recipes, too many of which contained opium, heroin, cocaine, or booze. His administration pushed for passage of the nation's first federal

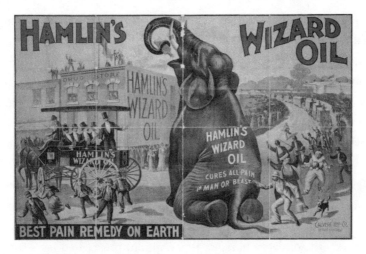

A patent medicine advertisement, c. 1890. Calvert Lithographing Co.
Lithographer. Hamlin's Wizard Oil. Courtesy: Library of Congress

drug-control legislation, the Pure Food and Drug Act of 1906 (over
the vigorous opposition of patent-medicine lobbyists).

He got his legislation. Most of the emphasis was on ensuring
untainted food, while the drug part of the act was relatively toothless,
little more than a set of regulations ensuring more accurate ads for
patent medicines. But TR was just getting started. He took on the
China opium trade, helping initiate the first International Opium
Conference in Shanghai in 1909 and strongly favoring a second in
the Hague two years later. In 1909 the United States passed a federal
Opium Exclusion Act, an important step in criminalizing the drug,
then signed the first international drug control treaty in 1912.

It was all capped by the nation's first significant antidrug law,
the Harrison Act of 1914, which regulated and taxed the production,
importation, and distribution of narcotics. What was a narcotic? Doctors used the term to describe drugs that produced sleep and stupor.
But to police and legislators, narcotics were heavy drugs, ones that
caused addiction. So the Harrison Act included cocaine by name,

despite the fact that it revs users up instead of putting them to sleep. Oddly, the first version did not include heroin by name (although it was added to the legislation years later). Mostly, the Harrison Act was targeted at opium and morphine. For the first time, all U.S. physicians and druggists had to register, pay a fee, and keep a record of every transaction involving opium, morphine, or cocaine. The act marked a watershed in the control of narcotics in the United States.

Patent medicine makers fought it, arguing that this was an infringement of Americans' long-standing right to decide for themselves what medicines to take. But they couldn't stop the regulation. After the Harrison Act passed, honest doctors, faced with having to keep records of every narcotic prescription, prescribed less. Druggists became far more cautious. Patients were more likely to think twice. Opium shipments into the United States plummeted from 42,000 tons in 1906 to 8,000 tons in 1934.

The stage was set for a question that is still being asked: Is drug addiction a moral failing or a disease? In other words, should drug addicts be treated as criminals or patients?

The Harrison Act sharpened the focus on this question, placing the government squarely on the side of criminalization. This left many physicians in a tough spot. A doctor could still prescribe and administer narcotics, but, the act read, "only in the course of his [sic] professional practice." Treating a patient's pain with morphine after surgery, for instance, was okay.

But what about treating a patient for morphine addiction? Was that allowable? Before the act, most doctors viewed drug addiction as a medical problem; their job was to cure it. They prescribed morphine or heroin to their addicted patients to help control the quality and lower the amount, gradually easing addicts off the drugs. But Harrison viewed narcotics addiction as a crime, not a disease, so using narcotics to treat it was not a legitimate professional practice. Therefore doctors who prescribed narcotics to addicts were themselves criminals. Bizarre but true: Within a few years of the Harrison Act, around 25,000 physicians were arraigned on narcotics charges; of those, some 3,000 were convicted and sent to jail.

Unable to get a legal dose, addicts, as always, turned to the streets. After the Harrison Act, the illicit drug market bloomed. It was the start of a long romance between crime and drugs. By 1930, about a third of all convicts serving time in U.S. prisons were there because they'd been indicted under the act.

The Harrison Act was reinterpreted in 1925 to allow some medical prescriptions for narcotics addicts, but by then the pattern had been set: Addiction to narcotics, in the eyes of the government, was a criminal activity. Opium addicts were no longer habitués, morphine addicts were no longer neighbors with a deplorable habit. Now they were junkies and hopheads driven mad by their yen for the drug (all terms that linked opium to the Chinese). The specter of Fu Manchu arose, along with a thousand other pulp images of leering Chinese men threatening innocent white women in smoky rooms. It was a cruel twist of history. British merchants pushing Indian opium had made addicts of millions of Chinese. Now the Chinese were the bad guys, while the heroes, like Fu Manchu's archenemy Nayland Smith, were British.

Ironically, one of the biggest beneficiaries of the Harrison Act was heroin. After Bayer stopped marketing it and legal availability shrank to nothing after 1914, heroin quickly became a street drug. It was relatively easy for criminals to make from morphine, or even from raw opium. And it was easier to hide and move around than liquid morphine. Heroin was made as a powder and was so concentrated that a few bricks were worth a fortune on the street. It was so powerful that it could be cut with other drugs or inert filler and sold to users in small, easy-to-hide packets. There were reports of "sniffing parties" where young people snorted heroin. There were stories of pathetic addicts dying in the back alleys of small towns.

By the time it was added by name to the Harrison Act in 1924, it was already an underground fashion among the young sheiks and flappers of the Jazz Age, especially popular in big cities like New York. And in Hollywood, where a 1920s dealer known as The Count became famous for putting heroin in peanut shells and selling it by the bag. One of his clients was Wallace Reid, famed as the world's most perfect lover and the handsomest man in pictures. As Reid's heroin addiction grew, his career hit the skids; he ended up dying in a sanatorium in 1923.

While the United States criminalized drugs, Great Britain took another path. In 1926 a select committee in London decided that addicts were medical patients, not criminals, an attitude that has shaped the practice of British medicine ever since. In the 1950s, for instance, dying patients in Britain could still get a Brompton cocktail, a potent mix of morphine, cocaine, cannabis, chloroform, gin, flavorings, and sweeteners. "It brings optimism where there is no hope, a certainty of recovery while death comes nearer," wrote one physician.

You might not be able to get a Brompton cocktail anymore, but Britain remains the only nation on earth where it is legal for a physician to prescribe heroin (although this is done rarely, usually for pain control in end-of-life care). And heroin addiction rates in Britain today are a fraction of what they are in the States.

HEROIN IS PART NATURAL—made from morphine, one of the naturally occurring alkaloids in opium—and part synthetic, the result of tinkering with the natural molecule, adding and subtracting atoms. It's what's called a "semisynthetic" opiate drug.

After 1900, many labs were doing what Bayer did to create its new semisynthetic heroin. They started with the alkaloids in opium—morphine, codeine, thebaine, and others—and tried to find out what made them work. These are not easy molecules to study. Morphine, for instance, has a complicated structure with five rings of atoms tied together. Some labs tried to strip it down to its smallest active component, breaking it into fragments, looking for the heart of the molecule. They then played with those fragments, substituting different atoms and adding side chains, making them into semisynthetics.

Around World War I, chemists searching for that holy grail of a nonaddictive painkiller made and tested hundreds of semisynthetic variations, few of which reached the market. But some of them were successful. Codeine was tweaked in 1920 to make hydrocodone (which, when mixed with acetaminophen, makes today's Vicodin). Doing something similar with morphine resulted in hydromorphone, patented in 1924 and still used today under the brand name Dilaudid. In 1916 chemists refashioned codeine to make oxycodone, a very strong semisynthetic known for being the key ingredient in Percocet (and now infamous in a potent time-release formulation under the name Oxycontin). These are all semisynthetic opiates, they're all effective painkillers, they all can make users a little spacey, and they're all addictive.

Others were found that were stunningly strong. In 1960, for example, a Scottish drug team was cranking out variation after variation of thebaine, another of the natural alkaloids found in opium. One day a worker in the lab used a glass rod sitting on a lab bench to stir some cups of tea. A few minutes after drinking it several scientists fell to the floor, unconscious. The rod had been contaminated with one of the new molecules they'd been working on. It turned out to be a super semisynthetic, thousands of times more powerful than morphine. Under the trade name Immobilon, it found use in darts to knock out elephants and rhinos.

The semisynthetic Oxycontin (aka oxy, cotton, kickers, beans, and hillbilly heroin) has made a lot of headlines as today's opiate du jour. The United States consumes about 80 percent of the world's supply. It has succeeded in moving opiate addiction from inner-city streets to middle-American small towns. It is everywhere, taken by just about every kind of citizen, but it's especially popular among poor, white, rural Americans. Overdoses (usually when taken with alcohol or other opioids) and oxy-assisted suicides are a major reason why the average lifespan of this group is declining—a downward shift that goes against everything medicine has done for the past century.

There's plenty of information out there about why oxy became so popular; all you have to do is read the news. But at the heart of it is the same simple fact that made China into a nation of addicts 170 years ago, made morphine a national scandal in the 1880s, and made heroin the most infamous drug of the 1950s. It's an opiate. And every opiate, without exception, is highly addictive.

After decades of work and thousands of failures, the semisynthetic path never did lead to that magic, nonaddictive molecule. So researchers took the next step, looking for a different approach: They sought a class of medicines not based on morphine or codeine or any part of opium at all, but something completely new. Something with an entirely new structure. Something completely synthetic.

Remarkably, they found some. The most powerful of these new synthetics—drugs like fentanyl and carfentanil—are not just as good as morphine at treating pain; they can be hundreds of times better. But they are also, and without exception, highly addictive.

The story of the synthetics, so important to understanding our current epidemic of opioid abuse and overdose, is found in chapter 8 (page 187).

CHAPTER 5

MAGIC BULLETS

PHYSICIANS IN THE YEARS just before World War II considered themselves thoroughly modern. They were masters of surgery. They knew—or thought they knew—all about bacteria's role in diseases. They had a growing number of effective vaccines. They were learning about all the major vitamins. They had access to sophisticated tools like pH meters and electron microscopes and X-ray machines and radioisotopes, and were using them to delve toward the very roots of disease. There was great optimism that ultimate answers would be found at the level of genes, proteins, and other molecules of life, and that scientists were about to break it all open. But in one basic way, medicine in 1930 was no more advanced than the healing practices of prehistoric humans. The white-coated modern physician was just as helpless as a rattle-shaking shaman when it came to curing most infectious diseases. Once a dangerous bacterial infection started in the body, nothing science could do could stop it. It either progressed and killed the patient, or the body fought it off on its own.

And bacteria were the cause of killer epidemics that swept through towns and across nations: pneumonia, cholera, diphtheria, tuberculosis, meningitis, and a hundred others. The vast majority of bacteria in nature are harmless to humans, or vital for health (you would die without the beneficial bacteria in your gut). But a few were dangerous. And those few could not be stopped.

Among the worst bacterial infections were those caused by a few strains of streptococci. These hardy bacteria are found everywhere—in dirt, dust, the human nose, on the skin, in the throat. Most are harmless. But a few are killers. Strep can cause more than a dozen different diseases, ranging from annoying rashes to strep throat to scarlet fever. One of the most dangerous is a strep blood infection. Before the 1930s, anything that got the wrong strep into your blood could be a disaster—and something as small as a nick with a dirty razor could do the trick. Once that happened, if the bacteria bloomed into a blood infection, all the money and power in the world could not save you.

In 1924 the teenage son of President Calvin Coolidge got a blister on one of his toes after playing tennis at the White House. He put a little iodine on it and forgot about it. But it got worse. By the time the White House physician was called in, it was already too late. The blister was infected with the wrong kind of strep, and the bacteria had gotten into the boy's bloodstream. He fought the infection for a week. But despite anything the best doctors in the nation could do, he died.

Strep was every doctor's nightmare.

WE TAKE ANTIBIOTICS for granted now. If our kid gets an ear infection, we give them an antibiotic. If a grandparent falls ill with pneumonia, they get an antibiotic. If our cough goes on too long, we ask about an antibiotic. These drugs have saved countless millions of lives—so many that experts credit antibiotics alone with extending the average human life span by ten years.

Ask most people what the first antibiotic was, and they'll answer penicillin. But the real antibiotic revolution started years before penicillin was widely available.

It began in Germany with a cage of pink mice. The cage was in a back room in one of the Bayer company's German laboratories. The year was 1929.

Bayer, now rich with cash from a string of drug discoveries, from heroin and aspirin to new sleeping medicines and heart drugs, had set its sights on solving the problem of bacterial infections. The path the company followed started with chemicals it was familiar with: fabric dyes. Bayer had started as a dye company. Now they were looking for dyes that could help cure disease.

The dyes-as-medicines approach—pioneered by the Nobel Prize–winning chemist Paul Ehrlich—made a lot of sense. Ehrlich knew that some dyes could stain certain animal tissues while ignoring others. Methylene blue, for instance, had a particular affinity for nerves. Dye a thin slice of muscle with methylene blue, put it under a microscope, and the nerves popped out of the background as a network of delicate blue fibers. The dye stained the nerves, not the muscle. Why was that?

Ehrlich was a maestro of dyes, discovering new ones, testing which ones liked to link to which tissues, trying to understand the reasons. He knew that some dyes also stuck to bacteria in preference to human cells, which led to a brilliant idea: Why not use these bacteria-specific dyes as weapons? What if you attached a poison to the dyes, made them into guided missiles that would attach specifically to bacteria and kill them, without doing anything to the human tissues around them? Could you cure a bacterial infection inside the body that way?

He called his idea for a new kind of medicine *Zauberkugeln,* magic balls. Today we use another term. Imagine a police detective chases a murderer into a packed theater lobby. The cop pulls out a gun and, without aiming, shoots it into the middle of the crowd. No worries: His gun is loaded with magic bullets that zip and dodge around the innocents and find their way to just a single target, the murderer, killing the culprit without harming anyone else in the room.

That's what Ehrlich envisioned: a drug that acted like a magic bullet. A medicine that would kill only the invader, leaving the patient whole. Today we call them "magic bullet drugs."

Ehrlich spent years trying to turn his inspiration into medicine. After making and testing hundreds of chemicals, enduring failure after failure, in 1909 he came up with a dye-based medicine that seemed to work, at least against one kind of bacteria. He named it Salvarsan. It was rough stuff, a dye-like core linked to arsenic as its poison, and it caused terrible side effects. But it worked to stop syphilis, a killer even more horrible than Ehrlich's medicine. Before Salvarsan there had been no cure for this increasingly common disease. Now there was a modern, high-tech cure that came out of a scientific laboratory.

Ehrlich's Salvarsan was not a very good magic bullet—it was too toxic for normal tissues and only worked against one disease—but it

Paul Ehrlich. Photograph, 1915.
Wellcome Collection

proved that a scientist could design a new chemical made to stop a bacterial infection, and that it could work. That was stunning.

And it led nowhere. Despite throwing himself into the search for more magic bullets, Ehrlich never found another. Neither did any other researchers through the 1910s and 1920s. Maybe Salvarsan had been a fluke. Most scientists gave up the search.

Bayer was one of the few companies to stick with this line of research. In the 1920s the German company went all-in on the hunt for another antibacterial. To do it, the company invested in and created something new: a large-scale, integrated process devoted to the creation, testing, and marketing of new synthetic drugs. Instead of relying on the hit-and-miss inspirations of an individual genius like Ehrlich, the Bayer labs would bring teams of technicians, modern corporate organization, and lots of money to the field, turning drug development into a factory operation—an assembly line for discovery. They would do for drugs what Henry Ford had done for cars in America.

Bayer already had teams of chemists looking for new dyes. These experts at tinkering with molecules were coming up with new substances all the time, most of them variations on synthetic dyes made from coal tar. Bayer chemists churned out hundreds of new chemicals every month. And almost none of them had been tested for medical use. No one knew what they could do. Maybe they had already created some powerful new medicine in their dye research and it was just gathering dust in a storeroom. Maybe they were sitting on a gold mine.

So Bayer decided to screen them all for use as medicines. Well, perhaps not all of them, but under the direction of a medical man, they could at least test a lot of them and then follow the most promising leads. Something new, something exciting, was bound to turn up. Even if there was just a hint of something positive, that hint could then be explored by chemists making new variations, playing

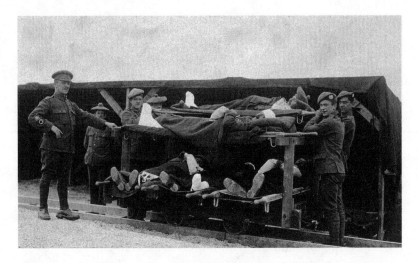

WWI: Pushvillers, France: wounded soldiers on a trolley. Wellcome Collection

with the molecule, tailoring it, teasing out more healing power. In the end they might come up with another aspirin or, even better, one of Ehrlich's magic bullets to fight bacterial infections.

The company had plenty of chemists, managers, and factory space. What it lacked was that medical man. So they hired a quiet young physician who was eager for the challenge. His name was Gerhard Domagk. And he turned out to be a brilliant choice.

Domagk had come of age serving in a German field hospital in World War I, sorting, stripping, and washing the wounded as they rattled in on wagons, occasionally helping with an operation. The men he treated had been ripped to shreds by new high-explosive shells and lacerated by machine-gun fire; many had been lying in the muck of trenches until they could be moved, so their wounds were deep, ragged, and filthy.

There, ministering to these mangled bodies, he saw something that changed his life. In countless cases it seemed they had saved the soldiers' lives: The skillful work of surgeons successfully repaired the wounds, then they were stitched up and sent to a recovery tent.

But there, a few days later, everything would go wrong. The wounds would turn red and start weeping, the first signs of an infection that would transform the carefully stitched tissue into festering, blackening, stinking sores. Postoperative wound infections like these killed armies of soldiers in World War I. They were caused by bacteria, that much was known, but no amount of cleaning and disinfecting seemed able to get out all the bacteria. It was often strep at the start, then it would turn to gas gangrene, with bacteria flooding into the bloodstream, releasing poisons, eating away the body as they advanced. Doctors would try to cut their way ahead of the infections, amputating and re-amputating limbs, trying to cut out the infection ahead of the bacterial advance. Too often they lost. Soldiers died by the hundreds of thousands—with more killed by wound infections during World War I, by some counts, than by bullets.

"I swore before God and myself to counter this destructive madness," Domagk wrote later. Finding a way to stop wound infections became his life goal. He went to medical school and spent a few years in a university lab as a medical researcher, where he proved himself solid and meticulous, and came up with promising ideas about fighting bacterial infections. But Domagk was denied advancement. He had a young family, and he couldn't see a way to make enough money to support them. Then Bayer reached out to him with an offer so good he couldn't turn it down: He was asked to take charge of a well-funded project to find new medicines. He would be given a larger salary, a new lab, and far greater responsibility. Among his targets would be the kinds of bacteria he fought during the war. In 1927, Domagk started working at the Bayer plant at Elberfeld.

Domagk's operation, a suite of the most modern labs, animal quarters, and offices, took up a third of a brand-new building. Into it flowed a stream of never-before-seen chemicals produced by Bayer's chemists. His job was to see if any of them had medical applications. So he figured out a way to screen the substances at an industrial scale, testing scores per month, hundreds per year. He focused on fighting bacterial infections, partly to pay back his comrades from the war and partly because that's where the money was. The biggest rewards would come from something that could beat the biggest diseases, and nothing was bigger than bacterial infections. A drug that could conquer tuberculosis or pneumonia—two of the biggest killers of the day—would generate huge profits.

All they had to do was find it. Domagk tested each new chemical two ways. The first was to mix it with disease-causing bacteria in a test tube to see if it would kill them. This was the less important of the two tests: plenty of chemicals, from bleach to pure alcohol, could kill bacteria in a test tube. That didn't mean they would make good medicines. The second, more important, screening was in live animals. These were usually mice (cheap, small, easy to breed in captivity), and, for the most promising candidates, rabbits. For a test, the mice would be separated into groups of six, each group in its own cage, each mouse injected with enough disease-causing bacteria to kill it within a few days—a shot of tuberculosis or pneumonia, or a particularly virulent strain of strep, and so forth. Then they were dosed with various dilutions of the test chemical (or an inert substance to act as a control); marked with colored ink depending on the disease, the chemical, and the dose; and then they were watched.

For years, all the mice in those cages died. Domagk's lab screened thousands of industrial chemicals. And stacks of lab notebooks recorded disappointing test results. Tens of thousands of mice were infected and killed. And not a single interesting medicine showed up. They tried dye after dye after dye. Nothing. They tried a series of gold-containing compounds. Nothing. They tried variations on quinine. Nothing.

Domagk's test system was working perfectly; he had created a flawless machine for uncovering new medicines. But where were

Portrait of Gerhard Domagk.
Wellcome Collection

the results? There were whispers: Looking for chemical cures was a waste of time. Living creatures were too complex, their metabolisms too unusual, for any industrial chemical to do anything. It was an expensive wild goose chase.

Domagk's bosses, however, kept the faith. All it would take was one patentable drug, one breakthrough, to earn back their investment. They kept going, pumping money into the process, waiting patiently for a blockbuster cure.

Finally, in the summer of 1931, it looked like they might have found one. Domagk's chief chemist, a high-strung, fabulously talented young researcher named Josef Klarer, had been working with a family of molecules called azo dyes, often used to color fabrics a vivid orange-red. It looked like some of these azos had a weak ability to kill disease-causing bacteria in mice. Once on the scent, Klarer spent months trying to create a stronger effect, tinkering with the azo dye core, trying to find stronger variations. About one hundred attempts later, he made a modification that enormously increased the bacteria-killing strength of the molecule. Thus inspired, he kept going, following it up with an even better variation that in some cases could completely cure strep infections in mice.

Domagk was elated. His bosses were elated.

Then everything went sideways. For some reason—no one knows exactly why—Klarer's azo dye variations stopped working. Instead of getting more powerful, each new molecule Klarer came up with seemed less effective than the one before. By the start of 1932, the trail had gone cold. The chemist tried every trick he knew, attaching various atoms at various places, trying to get the power back. Nothing worked.

This was not supposed to happen. Domagk's system was supposed to eliminate just this sort of random reversal. It was supposed to make the process more scientific, less chancy. Klarer had given them a glimpse of success, and it had been snatched away. What had happened?

Months passed as Klarer searched for an answer. He created scores of new azo dyes. All were failures. Then, nearing exhaustion in the fall of 1932, he tried one more. This time he attached a common, sulfur-containing side chain to the azo dye core. The side chain was nothing special, an industrial chemical that had been used in dye-making for decades to help make colors stick more tightly to wool. It sat on the shelves of every dye-making company in Germany. It was called sulfanilamide. But everyone just called it "sulfa."

THE BREAKTHROUGH HAPPENED while Domagk was on vacation. He was glad to get out of town in the fall of 1932, away from both months of failure in the lab and the national news, which focused on a right-wing fringe group that was about to take power.

They were led by a former soldier and mesmerizing speaker named Adolf Hitler, whose accession to the chancellorship was just a few weeks away when Domagk took his break from work.

While he was gone, his lab functioned as usual, screening chemicals against bacteria. One of the test chemicals was Klarer's sulfa-containing azo dye. The women who performed the tests on mice—Domagk's assistants for animal testing were almost all women—ran the tests as usual. Their job was to watch over test animals infected with some of the worst diseases on earth; they were accustomed, in the end, to seeing cages full of dead mice. But this time they found cages full of survivors, as one of them later said, "jumping up and down, very lively." When Domagk returned from vacation, his assistants proudly presented him with a large chart of the results. "You will be famous," one of them told him.

Domagk wasn't so sure. The results were too good. There could have been some sort of mistake. He immediately retested Klarer's new molecule, and retested again. And then again. The numbers they were getting were like nothing Domagk had ever seen. They were like nothing anyone had ever seen.

The sulfa-linked chemical completely protected mice from strep infections. It worked when taken by injection. It worked when taken by mouth. It worked at every dose level and appeared to work without any serious side effects (the worst thing that happened was that the red-colored medicine dyed the skin of the mice pink for a short time). It didn't work well on every type of bacteria, but it worked flawlessly on strep. When his team saw the cages of healthy mice, "We stood there astounded," Domagk remembered, "as if we had suffered an electric shock."

Domagk's bosses at Bayer were ecstatic. After five years of failure, their gamble was finally paying off. That sulfa side chain that Klarer tacked on looked like a key that switched on the bacteria-killing power of azo dyes.

For Klarer, this was just a starting point. He now focused on sulfa-containing variations, moving pieces here and there, zeroing in on even more powerful versions. By late November he had the best yet, a dark red azo dye the company called Streptozon.

Bayer quickly applied for a patent for the new miracle drug and leaked out samples to a few local physicians for tests on patients. The doctors were astounded by the medicine's ability to quickly cure patients who looked like they were at death's door. A few of them made presentations at their local medical societies, doctors talked to doctors, and word spread as far as France and England that, as one researcher put it, "Something was brewing in the Rhineland." And then, mysteriously, Bayer went silent about its new drug. There was no grand announcement. There were no scientific papers. No news stories. And no sales.

Two years would pass before Domagk published his first scientific paper on the discovery, and only after that did Bayer start pushing sales of Streptozon under a new trade name, Prontosil.

Why the long wait? It's a complex story, but at the center of it was a single problem: Soon after they got early samples of the new dark red and much talked-about miracle drug, researchers in France discovered that the power of Bayer's drug came not from its red azo dye, as the Germans thought, but from the little side chain Klarer had tacked on. The patient's body, once the drug had been taken, broke it into two pieces. The dye part did nothing but turn skin pink. The sulfa, a white powder that had first been made decades earlier, did all the work. As one scientific wit of the day put it, "The German's complicated red car had a simple white engine."

The problem was that sulfa, that simple white engine, couldn't be patented—it had been around too long, its original patent expired; it was cheap, easy to make, and available in bulk. Crates of the great miracle drug had been sitting in warehouses for years. Given that, who would pay a premium for Bayer's carefully patented red dye version? It appears that the company had gone silent for those two years because it couldn't figure out how to make money off it. During those many months, sulfa might have saved thousands of lives. But it seems that drug companies, like the drugs themselves, are not all good or all bad. They are both.

Just then—after Domagk's first scientific report of Prontosil's power had been published, but before it was widely available—fate gave the German's red dye a boost. And fate, as it often does, came in disguise—in this case, the disguise of a rich couple dressed up as German peasants.

THEY WERE AMERICA'S dream couple. He was Franklin Delano Roosevelt Jr., a tall, strapping Harvard student and the eldest son of the president of the United States. She was Ethel du Pont, one of the richest and most attractive young socialites of her day, heir to a chunk of an enormous du Pont family fortune, made from the manufacture of gunpowder and chemicals. The nation's newspapers couldn't get enough of them; they were followed by popping flash-bulbs wherever they went, trailing a wake of society page notes about every sporting event and play they attended, every elegant party they danced through.

There was, for instance, the November 1936 party at the Hock Popo Ski Club at the Agawam Hunt Club. That night at the Rhode Island club it was as if the Great Depression did not exist. The ball-room was packed with plutocrats and politicians, celebrities and members of the local power structure, dressed in the most ridiculous outfits. It was a costume party. Franklin Jr. went full-out German peasant in lederhosen, a bolero jacket, and a Tyrolean hat with a feather. Ethel complemented him in a dirndl skirt, a straw hat, and a blouse trimmed with edelweiss. It was a strange choice given the

Roosevelt administration's growing concern about Hitler and his Nazi Party.

But that, it turned out, wasn't the important thing. The important thing was that Franklin Jr. had a sore throat and a slight cough, not bad enough to leave the party early—there was drinking until the early hours—but enough to regret the party the next day. The throat got worse. A few days later a fever put the young man in bed. Just before Thanksgiving, he was admitted to Massachusetts General Hospital in Boston with an acute sinus infection.

No big deal, the doctors thought. A few days of bed rest and something to take the fever down would set him right.

In 1936, the art of medicine was well on its way to becoming a science. Two centuries' worth of advances in anatomy, physiology, pharmacology, and a dozen other fields had uncovered the workings of the human body and the things that could go wrong with it. Now a new field called "molecular biology," a more detailed understanding of life at the level of proteins and genes, was getting underway. Frock-coated doctors who performed surgery with their bare hands had been replaced by lab-coated technicians of health who did their work in gleaming modern hospitals. This was an age of science, hygiene, and medicine that actually worked.

Except that there was almost nothing that could help FDR Jr.

Instead of clearing up as expected, FDR Jr.'s sinus infection got worse, keeping him in the hospital. His mother, Eleanor Roosevelt, became so alarmed that she insisted on hiring a new doctor to take over his care—a top ear, nose, and throat man who immediately became worried that the president's son was in much more danger than anyone thought. There was a tender spot forming under the young man's right cheek that looked like the beginnings of an abscess, a pocket of infection. When he sampled the bacteria causing the abscess, he uncovered one of the most dangerous strains of strep, one that could release poisons as well as mount deadly blood infections. If those germs broke out of the abscess and got into the bloodstream, chances were good that the president's son would die.

The doctor decided to gamble. He had read reports in German medical journals about Bayer's new experimental drug, the red one, that had been shown to work particularly well in fighting strep infections. The results they were getting in Germany were close to miraculous; he knew that the drug was being tested at Johns Hopkins and had its enthusiasts there. Would Mrs. Roosevelt be willing to let him try it on her son?

Making a guinea pig out of the president's eldest son was not an appealing option. But after studying the question for a day or two as FDR Jr.'s condition worsened, the First Lady gave her approval.

In mid-December, FDR Jr.'s third week in the hospital, his fever raged and the infection worsened. His physician gave him the first injection of the new German drug, a dark red liquid called Prontosil that was shipped to the United States in carefully packed glass vials. The doctor, once he received the drug, wasn't sure how much to give his patient. The drug was too new, too little used, to know proper dosages. So he gave FDR Jr. what he thought was probably a really good dose, a lot of drug, watched for effects, then woke the young man every hour to give him more. Ethel du Pont was at his bedside. Eleanor Roosevelt sat on a chair outside his room, answering correspondence as the hours ticked by. The long night passed with little change. Then, the next day, FDR Jr.'s fever began to go down. It looked like the swelling around the abscess might be shrinking. The patient was sleeping more soundly, and he had more energy when he was awake. Later that day, the fever went away completely. The physicians observing his case were amazed. They had never seen a strep case with such a sudden turnaround.

A few days after Christmas, FDR Jr. was released. The strep was gone. He would later marry Ethel du Pont (the first of five marriages), be decorated for his service in WWII, and serve three terms in Congress. But of all these accomplishments, perhaps the most important was that he had been the first American to demonstrate the power of the world's first antibiotic.

Word of his miraculous recovery, trumpeted in every newspaper in the country, spurred a sulfa craze. Everyone started demanding it.

And as soon as drug companies realized that the active component in Prontosil, the "little white engine" sulfa, was out of patent, they all started making sulfa-containing medicines. Pure sulfa alone worked; the little white pills were cheap and effective against anything caused by strep. But with a bit more research, drug chemists found that by attaching the sulfa side chain to different molecules, they could make versions that worked against different bacteria. Prontosil could stop strep blood infections, scarlet fever, gas gangrene, erysipelas, cellulitis, and childbed fever. New formulations expanded sulfa's effectiveness to other major diseases, like pneumonia, meningitis, and gonorrhea. And these new versions could be patented. "The most sensationally valuable new medicine in many years," trumpeted the *New York Times*. "A modern miracle," headlined *Collier's* magazine.

Overenthusiastic doctors started using it for everything. The joke at one hospital was that when a patient came in they were immediately given sulfa, and if they weren't better in a week they might get a physical exam. It was available without a prescription, so nurses made their rounds with a handful of pills in their pockets, passing it out like aspirin. It cost almost nothing, had few side effects, and seemed good for just about anything that ailed you. By the fall of 1937, American drug firms were making more than ten tons of sulfa drugs per week.

The new drug's honeymoon was hot, bright, and brief. No effective drug is without side effects, and as use spread, sulfa's started showing up. Pure sulfa right out of the can was still remarkably nontoxic, with just a few serious problems centering around rare allergic reactions. But the American Medical Association, watching the drug's rapid spread with concern, warned that one or another of the fast-growing number of new sulfa variations might prove more toxic, and that not much testing had been done on most of them.

The AMA was right.

IN THE FALL of 1937, kids started dying in Tulsa. They showed up first in doctors' offices complaining of terrible stomachaches, then stopped urinating, then fell into comas, and within a short time six children were dead. And more kept showing up.

It was a mystery that took local health authorities a few weeks to solve. The common thread in the illness was a new medicine called Elixir Sulfanilamide, a sweet, liquid form of the drug made by a patent medicine company, Massengill. The company's idea was to put sulfa into something that would appeal to kids, women, and the black community, all of whom were thought to prefer sweet liquids to bitter pills. Now it looked like Elixir might be killing people.

Doctors in Tulsa contacted the American Medical Association, and word was passed to a new and very small federal operation called the Food and Drug Administration (FDA), which sent one of their few agents to Tulsa to investigate. He found a bigger disaster in the making, with the local hospitals seeing more and more cases. Soon he had a pretty strong suspicion that it was Elixir that was to blame, and he faced a looming question: Where else was Elixir being sold?

It turned out that the drug had been on the market for a month and was being sold across the country. Massengill was assuring everyone that their medicine couldn't be to blame. But the AMA did tests on it and found out that the company had used a poisonous liquid, diethylene glycol, a common ingredient in antifreeze, to dissolve the sulfa.

While the AMA and FDA did their work, the death toll kept

rising. Some 240 gallons of Elixir had been distributed from the factory to salesmen to local drugstores to doctors and patients, much of it across impoverished areas in the South, where record-keeping was poor and the trail of the drug hard to track. Doctors were afraid of losing their licenses if they admitted they had recommended it. Druggists were loath to admit they had doled out a poison. Buyers, like those who had bought it for gonorrhea, sometimes gave false names when they bought it. Massengill was still claiming it wasn't to blame. By mid-October the death toll was thirteen.

A typical case involved a druggist in Georgia who bought a gallon of Elixir, then doled it out to patients in smaller bottles. He told the FDA that he had only sold six ounces. But when they measured the remaining liquid, they found that twice as much was missing. They confronted the druggist; he admitted that he had made two additional sales. Both of the purchasers were dead.

The newspapers got hold of the story, and alarm began to spread. By the time the Department of Agriculture (which oversaw the FDA at the time) made a report to Congress at the end of November, seventy-three deaths from the poison had been confirmed, plus one more: the head chemist of the Massengill company who, when he realized what he had done, shot himself.

It was the largest mass poisoning in American history. It was a national scandal. And it spurred something good: passage of the 1938 Federal Food, Drug, and Cosmetic Act, the first law in U.S. history that required that new drugs be proved safe before they went on the market, and that all active ingredients had to be listed on labels. The new law created the modern FDA. Much amended and expanded, it is still the foundation of today's pharmaceutical laws.

ANYBODY WHO HAS ever watched a World War II movie has probably seen that tense moment when a medic frantically shakes a whitish powder over a soldier's wound. That powder was sulfa. Mountains of the drug were used during the war to prevent the sort of horrific infections Gerhard Domagk had seen in his early days. U.S. companies made more than 4,500 tons of sulfa in 1943, enough to treat more than one hundred million patients; the Germans, helped in part by continuing research from Domagk, made thousands of tons more. And it worked. Deaths from wound infections in the Second World War were a small fraction of what they had been in the first.

Domagk's dream of fighting the "madness" of wound infections had come true.

IN 1939, DOMAGK was awarded the Nobel Prize in Physiology or Medicine. Unfortunately, he couldn't accept it. Hitler, angered by a Nobel committee's 1935 decision to award its Peace Prize to an anti-Nazi activist, had decreed that no German was henceforth to accept any Nobel. Domagk, a good German, didn't formally accept his own, but he made the mistake of writing a letter thanking the Swedish committee for the honor. A short while later the Gestapo showed up, searched his home, arrested him, and threw him in jail.

Later, he tried to make light of it, telling a joke about his time behind bars. "A man arrived to clean out my cell and asked me what I was doing there," Domagk would say. "When I told him I was in jail because I had received the Nobel Prize, he tapped his head and declared, 'This one is mad.'"

After a week the government felt it had made its point and set Domagk free. But he was a changed man. "It is easier to destroy thousands of human lives than to save a single one," he wrote in his journal. He was allowed to continue his research, but only after signing a curt letter to the Nobel committee refusing his prize. He began to suffer from anxiety and heart problems.

He continued working with sulfa, making new variations, extending its use to new diseases. It became a staple of Nazi military hospitals, as it was for the Allies.

It was the best medicine military doctors had until the very end of the war when, thanks to sulfa, something even better came along.

AROUND THE TIME Domagk was first recruited to find new drugs at Bayer, a Scot working in a London laboratory noticed something odd. In 1928 Alexander Fleming was growing bacteria on plates of nutrients and was unhappy to find a stray mold contaminating his samples. But there was something odd about this mold. Wherever it grew, it was surrounded by a clear, germ-free area, a sort of no-go zone for bacteria. It looked as if the mold were giving off something that somehow stopped the bacteria. Fleming tried to purify the active substance, doing tests on what he called "mould broth," which we now know as penicillin. But the active principle proved so hard to isolate and keep fresh that he eventually dropped the project. He turned his attention instead, like so many other scientists of the day, to sulfa.

Sulfa's success led other researchers back to the search for other magic bullet medicines, including Fleming's penicillin. During the war, energized by the need to find something that would work on even more types of bacteria than sulfa did, scientists figured out how to purify, produce, and store penicillin in large quantities. Once it was widely available in the waning days of the war, the new drug quickly pushed sulfa aside: Penicillin was more effective against more types of bacteria and better able to fight diseases, like syphilis and anthrax, that sulfa couldn't touch. Other bacteria-fighting chemicals were soon found in other molds and fungi: streptomycin, neomycin, tetracycline, and a score of others.

Advertisement for penicillin production from *Life* magazine. Science Museum, London

The age of antibiotics had started. By the end of the 1950s, antibiotics were being used to control just about every important bacterial disease. Epidemics that once routinely killed hundreds of thousands every year became things of the past. During the two decades after World War II, the death rate from childhood diseases dropped by more than 90 percent, and the average life expectancy in the United States increased by more than ten years. Demographers call this drug-produced sea change "the great mortality transition."

Sulfa kicked it off. Unlike other antibiotics made by living organisms, sulfa was created in a lab. But it accomplished the same purpose—selectively killing bacteria while leaving the human body alone, acting like one of Ehrlich's magic bullets—and it renewed medical interest in finding more such drugs.

And sulfa did more than that. It also pointed the way to a new system for finding more, and more powerful, drugs. Bayer's big-money corporate approach cements its status as one of the first modern drug companies. The credit goes to the company's long-term thinking and willingness to gamble, Klarer's brilliant tinkering with molecules, Domagk's efficient system for testing, and the building of a linked system of dedicated research labs and animal testing facilities under the direction of medical experts. This was a blueprint for today's pharmaceutical giants.

No longer would drug discovery be done by lone geniuses working on hunches. It would be done by scientific teams working on targeted problems, using chemical structure as a guide. Drug discovery would evolve from an art to an industrial science.

Sulfa changed not only the way drugs are discovered, but also the regulations that ensure their safety. The Elixir mass poisoning and the 1938 act that created the modern FDA laid the groundwork for today's legal system to make sure drugs are more or less safe and effective, plus how they're labeled. The 1938 U.S. legislation served as a model for the rest of the world.

Those achievements alone would make sulfa one of history's most important drugs. But the drug Klarer found and Domagk proved effective did something more, at a deeper level. Sulfa and the antibiotics that followed gave the public an enormous faith in pharmaceuticals. Drugs, it seemed, truly were miracles. Drugs could be found that could cure anything and everything, not only sniffles and headaches but humanity's most deadly diseases. Before sulfa, drugs were relatively weak, mostly palliative, limited in reach, and doled out at the corner drugstore without a prescription. Very few drugs could cure anything. That all changed after FDR Jr.'s miraculous recovery. In the drug-fueled optimism that reigned after sulfa and the antibiotics, it seemed like humanity might find drugs that could cure *everything*.

But the news was not all good. Antibiotics work against bacterial infections, but generally not viruses (vaccines are still the best things we have for avoiding viral diseases) or parasites (very different bugs that cause diseases like malaria; we are still searching for a game-changing antimalarial drug). So their reach is limited.

More important, perhaps, is the fact that the targets they do work against, disease-causing bacteria, are very good at finding ways to fight back. Some of them can create chemicals to neutralize antibiotics, others find ways to disguise themselves, and when they find an effective defense, they're often very good at passing it along to other bacteria, even those they're not closely related to. The process is called "antibiotic resistance." And here, too, sulfa was first.

Doctors initially noticed it among soldiers, many of whom were given sulfa just before going on leave as a way to prevent gonorrhea. If they got the clap, they were given more sulfa when they returned to duty. It worked great. In the late 1930s, the drug stopped gonorrhea more than 90 percent of the time. But by 1942, the rate was down to 75 percent, and it kept falling. The German army was having the same problem, often caused by soldiers taking just enough of the drug to make their symptoms go away, then stopping it before the bacteria were all gone. The few bacteria that survived were those most resistant to sulfa. They repopulated and spread. Sulfa resistance began building in strep cases as well; by 1945 one huge set of U.S. Navy tests of sulfa as a way to prevent strep infections was stopped because too many soldiers were getting sick. Sulfa was losing its kick as bacteria found ways to defeat it.

These early warning signs were ignored, however, in the general euphoria that accompanied penicillin and other antibiotics. If one antibiotic stopped working, patients were simply shifted onto another—until resistance started showing up there as well. Today, antibiotic resistance is a huge problem, capped by the appearance of a handful of bacteria that are resistant to all common antibiotics. Doctors are wisely prescribing antibiotics less often, and monitoring their use more carefully. More critical attention is being paid to the widespread employment of antibiotics to prevent disease and spur the growth of farm animals. We're still learning the lesson that overuse and misuse of these wonder drugs can carry a heavy penalty.

And sulfa? Sulfa is still around—various forms are used to treat ear infections, urinary tract infections, and other diseases—and has had something of a recent revival because of antibiotic resistance. Because sulfa was old-fashioned by the 1950s, it was used less and less. As a result, resistance to it dropped. So it often still works and, used carefully, is still a valuable tool in fighting infections—although now it is just a middling antibiotic among more than one hundred on the market.

CHAPTER 6

THE LEAST EXPLORED TERRITORY ON THE PLANET

Sirocco

Henri Laborit broke the surface and gasped for air. He had nearly been taken down, the *Sirocco* pulling him under until he fought free and kicked up through the dark. He was one of the lucky few with a life jacket. The sea was being churned by panicked men. Oil fires lit the surface. He had to fight off three soldiers, "unfortunate idiots," he called them, who, it appeared, did not know how to swim. They were panicking, arms windmilling, clutching at anything that floated, trying to use Laborit as a life raft. "I had to get rid of them," he wrote later, although he never wrote how he did it. Laborit kept his distance from the dying men, fires, and bobbing bodies, turned on his back—a swimmer's trick—and looked at the stars.

It was just after one in the morning the night of May 30, 1940. Henri Laborit was a junior medical officer on the small French destroyer *Sirocco*. They had been helping with the great evacuation of troops at Dunkirk, after the Nazi Blitzkrieg had smashed three Allied armies and trapped the survivors in a little pocket around the port, their backs to the English Channel. Every Allied ship within a day's steaming had raced to the area to pull men out of France. Laborit's destroyer arrived at the height of the action, zigzagging toward shore through clouds of black smoke and the remains of half-sunken ships. Soldiers lined the seawalls and beaches; some stood hip-deep in the water, rifles raised above their heads. The Germans were trying to kill everything that moved. "There were no doubts in the minds of the crew that their days were numbered," Laborit remembered. But the *Sirocco* managed to pick up eight hundred French riflemen, packed them shoulder to shoulder on deck, and, as dusk fell, pulled away. All they had to do now was make it to England.

Dover was less than fifty miles away, but the waters off Dunkirk were shallow and treacherous, and German planes were everywhere, so they moved slowly along the coast for miles, waiting for dusk and the chance to make a break for it. Everyone was on high alert. Around midnight, just as they were ready to dash for England, someone spotted a German torpedo boat emerging from behind a buoy. Laborit watched as two torpedoes sped toward them and barely missed, just off the bow, wakes shining in the dark. Then a second round of torpedoes hit them dead-on. The *Sirocco* shuddered; Laborit felt the stern lift. German dive-bombers zeroed in on the flames, and a second tremendous explosion ripped the *Sirocco* open. Laborit thought the ship's ammunition magazine had been hit. He saw riflemen's bodies flying through the air. Then he was in the water.

The ship sank quickly, the bombers flew off to find other prey, and Laborit floated on his back. As the hours went by, he saw men around him slowly losing their strength. He was very cold; his mind began to wander. He had been trained as a physician just before the war and knew what was happening. The freezing sea was sucking the heat out of his body, and hypothermia was setting in. If it went on much longer, he would die. How long? His fingers and feet were already starting to go numb, legs getting sluggish. Lower the body's temperature enough and you went into a sort of shock reaction, blood pressure plummeting, breathing growing shallow, the body going white and still. Would it take an hour? Several?

Laborit saw it happening around him. Ninety percent of the riflemen they picked up at Dunkirk would die that night. So would half of the *Sirocco*'s crew.

He forced himself to keep moving. He noticed that he was still wearing his helmet, a stupid thing to do, and fumbled with the strap until he got it off. He watched it slowly fill with seawater. There must be a hole in it, he thought. He stared at the helmet until it sank. His mind was slowing down.

Somehow he made it until dawn, when he saw a few dim lights and heard some distant calls. A small British warship was looking for survivors. He could see men, using the last of their strength, thrashing desperately around it, clamoring to get aboard. Sailors on deck were throwing lines, and the swimmers were clawing at one another to get to them. It was a madhouse. The *Sirocco*'s survivors were so weak that some could not make the climb—they would get partway up, lose their grip, and fall back on top of the others. Men were drowning. Laborit forced himself to hold off, waiting for the chaos to settle. Then, with an enormous effort, he swam alongside, grabbed a slippery rope, and began hauling himself up. He made the rail, was pulled over, and immediately passed out. He came to in a tub of warm water with someone slapping his face, saying, "C'mon, Doc, put some effort into it!"

LABORIT, SUFFERING FROM exhaustion and exposure, was taken to a French military hospital where, as he recovered, he fell into an odd, floaty sort of depression. Today we would call it post-traumatic stress. All Laborit knew was that he felt off-balance, as if the solid ground beneath him had turned to quicksand. "I found myself distraught by the idea of having to continue to live," he remembered. He was twenty-six years old.

But here, too, he fought his way back. There was some public attention to distract him. He was one of the heroes of the *Sirocco*, as the press called them. He was awarded a medal. He found comfort in working at medical duties. He developed a dark sense of humor. But still everything felt a little distant, as if he were watching life through a window.

When the French military felt he had recovered enough, his commanders decided that he would benefit from a change of scene and sent him to a naval base in Dakar, the capital of Senegal in North Africa. Here, in the sun and sand, he practiced general medicine for a few hours in the mornings and spent the rest of the day painting, writing, and riding horses. He had a slight build but was good-looking, almost movie star handsome with his thick, dark hair; he was also smart, ambitious, and accustomed to money—his father was a physician, his mother came from an aristocratic family—and a little snobby. He hated being exiled with his wife and small children in the sweltering "backwater" of Africa and desperately wanted to get back to France. To take his mind off his boredom, he decided to specialize in surgery. He found mentors among the doctors of Dakar and essentially trained himself in the arts of cutting and sewing, using cadavers from the local morgue. He had skillful hands. But he had little patience.

Once he started operating on living patients, despite his skill and best efforts, things often went wrong. Regularly, seemingly for no reason, in the middle of an operation a wounded soldier's blood pressure would nose-dive, his breath would grow shallow, and his heart would start racing. It was a bad sign. Often the patient died on the table, not from the operation itself, but from a condition called "surgical shock." No one knew what caused it, and at the time there was little that could be done to counter it. No one knew why some patients went into shock while others did not. Nothing seemed to alter the odds.

Laborit decided to find some answers. As he moved from post to post through the rest of the war, he sought out whatever medical literature he could find on the subject of surgical shock. And he began to put together a picture. Shock, most experts thought, was a response to injury (including the injury of being laid out on a table and cut open by a surgeon). Researchers were just learning that wounded animals react to injury by releasing a flood of chemicals into their blood, molecules like adrenaline that trigger a flight-fight-or-freeze response. Adrenaline increases heart rate, speeds up metabolism, and alters blood flow. Laborit came to believe that the keys to surgical shock would be found in the chemicals the body released into the blood when it was injured.

That was one approach, but not the only one. Some researchers thought shock was more mental than physical. After all, a shock response could be triggered by fear as well as injury. Threaten someone with a knife—convince them you're going to hurt them—and their heart speeds up, their breath grows ragged, they sweat. Mental stress by itself, in other words, can cause a shock response. Laborit had seen that with his own patients, who, in the hours before an operation, sometimes got so tense, so anxious about the coming pain, that they started exhibiting signs of shock well before scalpel touched skin. Maybe surgical shock was simply an extension of this—a natural reaction that went too far, that somehow spiraled out of control.

Laborit brought the two ideas together. His thinking went like this: The patient's anxiety and fear of pain before an operation triggered the release of chemicals in the blood. Then the physical shock of the operation ramped it up a notch. The mental stress and physical reactions were tied.

So perhaps the answer was to short-circuit the process by easing the fear before the operation. Ease the fear, lower the anxiety, and you might be able to block or slow the blood chemicals from triggering fatal shock.

But what were these chemicals? Very little was known about molecules like adrenaline because they were released in very small amounts, were quickly diluted to almost undetectable levels in the blood, and then disappeared entirely within minutes. More was being learned about adrenaline all the time, but it wasn't the only such molecule—there were others yet to be identified. Laborit read everything he could on the subject, thereby becoming one of those rare surgeons who deeply understood biochemistry and pharmacology, and began to play with ideas about moderating stress chemicals in the body.

His patients became his test subjects. When the war ended, Laborit was still in North Africa. But now his ennui was lifted because he was engaged in his research, testing ways to try to calm his patients and put them at ease before their operations. He would mix various drugs into chemical cocktails designed to lower anxiety. It was hard to find the right mix of ingredients. Physicians in the past had tried a lot of things to keep patients quiet, from shots of whiskey to sleeping pills, morphine to knockout drops (see page 81). But all of these, in Laborit's view, were imperfect. All of them had side effects, some of which could be dangerous. They weakened patients as well as relaxed them. They put patients to sleep. Laborit wanted his patients strong and tranquil, unworried before their operations, but not unconscious until they were actually on the table. The Greeks had a word for what he wanted: *ataraxia*, the mental state of a person who was free from stress and anxiety but at the same time strong and virtuous. He wanted to create *ataraxia* with drugs. So he kept searching and testing.

To this he added another idea, perhaps spurred by his time in the water after the *Sirocco* sank. He decided to try cooling his patients down. If he could slow their metabolisms, he thought, he might be able to help relax the shock reaction. He pioneered a procedure he called "artificial hibernation," using ice to cool his patients along with the drugs.

This approach, a historian later wrote, was frankly revolutionary. Other researchers were going the opposite way, trying to counter shock once it started by giving shots of adrenaline—exactly the wrong thing, Laborit thought. He was convinced that his artificial hibernation, coupled with the right drugs, would do the job.

RP-4560

By 1950, Laborit was reporting a string of positive results in medical journals. His work gained him so much attention that his bosses decided to rescue him from the hinterlands, bringing him to the center of all things French, Paris.

Ah, Paris! Paris was the world to any ambitious Frenchman (or -woman). It was home of the nation's political leaders and business headquarters; its religious elite and military top brass; its best writers, composers, and artists; the nation's top university (the Sorbonne) and leading intellectuals (the French Academy); its finest houses and most beautiful music, fashion, and food; the best libraries and research centers, museums, and training centers. If you were French and you were a leader in your field, your heart yearned for a post in Paris.

And now Laborit had arrived. He was transferred to the nation's most prestigious military hospital, the Val-de-Grâce, just a few blocks from the Sorbonne. There, with access to a wide range of experts and far greater resources, he widened his research.

He needed a drug expert, and he found one in the form of an enthusiastic researcher named Pierre Huguenard. Laborit and Huguenard went to work perfecting his artificial hibernation technique, accompanied by cocktails shaken up from atropine, procaine, curare, different opioids, and sleeping drugs.

Another of the chemicals the body released in response to injury, histamine, caught their interest. Histamines were involved in all sorts of things in the body, not only released in response to injury but also involved in allergic reactions, motion sickness, and stress. Perhaps histamine played a role in the shock reaction. So Laborit threw another ingredient into his cocktail: an antihistamine, a new kind of drug then being intensively developed for the treatment of allergies. And that's when things began to get interesting.

ANTIHISTAMINES LOOKED LIKE the next great family of miracle drugs. They had effects on everything from hay fever to seasickness, the common cold to Parkinson's disease. Drug companies were working feverishly trying to sort it all out and make versions that could be patented.

But, like all drugs, they also had side effects. One was particularly troubling when it came to marketing: Antihistamines often caused what one observer called "a disturbing drowsiness" (today's nondrowsy antihistamines were still decades away). This was not like the sleepiness caused by sedatives and sleeping pills. Antihistamines didn't slow down everything in the body. Instead, they seemed to be aimed toward a specific part of the nervous system: what doctors in the 1940s referred to as the sympathetic and parasympathetic nerves (today it's the autonomic nervous system). These make up the body's background nervous system, the signals and responses that operate below the level of our conscious minds; they are the nerves that help regulate breathing, digestion, and heartbeat, for example. And it would be among these nerves, Laborit thought, that the secrets of the shock reaction would be found. He wanted a drug that acted on these nerves specifically, without greatly affecting the conscious mind. Antihistamines looked like they might be just the thing.

So he and Huguenard began testing. They found that adding the right dose of the right antihistamine in the hours before surgery yielded patients who, although still conscious, Laborit wrote, "felt no pain, no anxiety, and often did not remember the operation." As an added benefit, Laborit found that his patients needed less morphine for pain. His antihistamine-enriched cocktails, along with artificial hibernation, were resulting in less surgical shock and fewer deaths.

But there was much more to do. He didn't really want an antihistamine in his cocktail—he wasn't treating allergies or motion sickness, after all—as much as he wanted one of the side effects of an antihistamine. He was looking only for the reduced anxiety, the euphoric quietude he had seen in some patients. He wanted an antihistamine that was *all* that side effect. So he wrote to France's biggest drug company, Rhône-Poulenc (RP), and asked the researchers there to find one.

Luckily, he reached the right people at the right time. RP was very actively looking for newer, better antihistamines and, like all drug companies, had plenty of failures sitting on the shelf—drugs that had been too toxic, or had too many side effects. They began retesting their failures.

A few months later, in the spring of 1951, the firm delivered to Laborit an experimental drug called RP-4560. They had stopped working with it because it had virtually no use as an antihistamine. But it had a strong effect on the nervous system. Animal tests had shown it was relatively safe. It might be just what Laborit was looking for.

It turned out to be the best thing he'd ever thrown into a cocktail. It was very powerful; a small amount was all that was needed.

And it did the job: RP-4560 given before a variety of different kinds of surgery, from wound treatment to minor operations, lowered patients' anxiety, improved mood, and lessened the need for other drugs. The patients who took it were awake and aware but seemed to tolerate their pain better and required less anesthesia in order to lose consciousness. It was strange, really. It wasn't that the pain was gone. They realized that they were in pain, but seemed not to worry about it. They knew they were going in for an operation, but seemed not to care. They were disinterested, Laborit found—"detached" from their stress.

His findings became the talk of Val-de-Grâce. And Laborit, now enthusiastic, became a promoter. One day over lunch in the staff canteen, he listened as a friend—the head of the hospital's psychiatric department—bemoaned the need to put his severely ill mental patients in straitjackets, an old lament raised by generations of caregivers for the insane. How sad it was that in many cases the mad were too agitated, too manic, too dangerous to be cared for without restraints. They would scream and thrash, sometimes attacking others or hurting themselves. So they had to be knocked out with drugs, or strapped into beds, or straitjacketed. What a pity.

It gave Laborit an idea. He told his luncheon companions that instead of restraints, they might want to try giving these manic patients a dose of RP-4560 and cooling them down with ice.

Bedlam

Every morning, the mad could be found in the waiting room at Sainte-Anne, where the dregs of the night before had been hauled in by police or family members, "brains boiling with fury, overwhelmed with anguish, or dead-beat," one physician recalled. They were the maniacs, the ravers, the ones who saw hallucinations and heard voices, the despondent, the lost.

When it became too much, they ended up at Sainte-Anne, the only psychiatric hospital within the boundaries of Paris. Every big city had its own version of Sainte-Anne, a government-sponsored mental hospital, an asylum designed to remove the mad from society, to help them and keep them safe—and out of sight.

They were called mental "asylums" for good reason: The mentally ill needed a refuge. For most of history, the mad had been left to the mercy of their families, who with rare exceptions hid the worst afflicted in back bedrooms and locked them into basements. Some were treated kindly, and others were chained, beaten, and starved.

That changed with the Industrial Revolution and the growth of cities. With increasing stress and the dispersion of families, the mad increasingly ended up on the streets. They became the responsibility of others—or of no one.

Charitable organizations were formed, and social movements mounted to care for them humanely. Beds, food, and medical care had to be found. In America, the answer in the nineteenth century was to build large asylums, designed to be models of advanced care, with park-like grounds, airy workshops, and professional therapy overseen by physicians trained specifically in the treatment of mental disorders. The asylum's design would allow separation of men from women, the violent from the nonviolent, the curable (who were often housed in the front, most visible rooms) from the incurable (often locked in the back, where the cries and smells would be less troubling to visitors). Diets would be healthful and simple, punishment would be rare, and here, as one writer put it, "they would gradually come to their senses thanks to the salutary effects of the asylum environment."

There would be benefits for medical science, too. With all sorts of madness gathered in one place, physicians of the mind would be better able to study a range of conditions under somewhat controlled

Man in a strait-jacket. French
asylum, 1838. Wellcome Collection

circumstances, allowing a deeper understanding of mental disease,
and increasing the chance of finding cures.

That was the ideal, in any case. And in many ways, it
was successful.

In Britain, for example, no more than a few thousand patients
were kept—imprisoned, often—in a handful of mental asylums
like the infamous Bethlem Royal Hospital, outside of London, better known as Bedlam. In the eighteenth century, Bedlam became
infamous for allowing bored visitors to pay a little money and walk
through to goggle at the inmates, turning madness into an evening's
entertainment. A century later there were sixteen large asylums in
the London area alone. The average number of patients per asylum
rose from less than sixty in 1820 to ten times that number within a
few decades. In America, numbers of patients rose just as fast. By

1900 American asylums were buckling under the strain of 150,000 mental patients.

Most of these were supported by the public, through state and county budgets, or by charitable organizations. The result was that the price of care in these public asylums was low for families. The numbers kept going up as more and more families dropped off their senile grandparents, alcoholic uncles, and mentally disabled children at bargain prices. Police did the same with drug addicts, corner ravers, and disturbers of the peace. Workhouses, poorhouses, hospitals, and jails added their own overflow. The huge asylums filled to bursting.

The Hospital of Bethlem [Bedlam] at Moorfields, London: seen from the north, with people walking in the foreground. Engraving. Wellcome Collection

Many of these inmates were curable. Asylums worked best for cases in which the patient had suffered a temporary mental breakdown or was working through a trauma and who, after a few weeks of rest and peace, could be released.

But many were deemed incurable cases. These included the "senile" elderly (today we would say they have some form of dementia, like Alzheimer's disease), the developmentally disabled, and those who had lost touch with reality entirely and could not find their way back. This latter group—the ones who curled into a corner and didn't move for months on end, or talked nonsense endlessly, saw things that weren't there, or heard voices telling them what to do—are now generally termed schizophrenic. Again, because no one was sure what caused any of these diseases, no one could fix them. As one expert put it, "In 1952, the six inches between one's ears were the least explored territory on the planet." What was known was this: Once these "incurables" entered an asylum, chances were they'd never leave. They were in the back rooms for life, joined by more every year. The overall numbers just kept going up, and the proportion of worst-case patients who could not be treated—only cared for—got bigger every year. By the early twentieth century almost every asylum was overstuffed and understaffed. They had morphed from places of rest and recuperation to loud, crowded holding pens, "loony bins" concerned less with curing than with security and sedation. Asylums became, as one expert said, "dustbins for hopeless cases."

In addition—and this turned out to be important—they were ever-deepening money-sinks for government budgets. The big asylums were, for the most part, funded with state and county tax monies, and as they grew, every year they took a bigger and bigger bite out of government budgets. Every attempt to cut costs led to less humane care. Reports of patient abuse grew. Taxpayers were getting tired of it.

What about the science? Here, too, nothing good seemed to be happening. The sad fact is that the chances of getting cured in a mental asylum in 1950 were not much better than they had been in 1880. There had been great excitement about the possibilities of better care through lobotomies and electroshock when these methods were first introduced in the early twentieth century, but after the enthusiasm waned, every new advance soon proved marginal. When it came to their toughest cases, especially schizophrenia, asylum doctors were making little headway. Psychiatrists, despite amassing an impressive and ever-growing knowledge base about mental health, were unable for the most part to help their sickest patients.

Powerless

The morning routine at Sainte-Anne mental hospital in Paris went like this in 1952:

The well-dressed heads of the hospital's main wards would visit the waiting area and review what the night had washed up at their door. The waiting area was a rich panoply of all that could go wrong with the human mind. Physicians could find examples of every sort of insanity, and noted those who fell into an area of current research interest. The morning review, one Sainte-Anne physician wrote, was like "Going shopping in the mental illness market."

The most interesting cases were marked down for those researchers who were known to have an interest in them. The less severe cases, those most likely to be helped, were taken to the Free Department for voluntary inpatients ("voluntary" was a misnomer; a few came in on their own, but most were brought by police or committed by family members). The tougher cases ended up in the more restrictive Men's or Women's Departments, the wards with locked doors, where they could be carefully monitored and, if necessary, restrained.

Somewhere else in the hospital on those mornings in the early 1950s, striding the halls or marching across the grounds trailed by a retinue of underlings, was the director of Sainte-Anne, Jean Delay. He was a short but commanding figure, a true intellectual in the mid-twentieth-century mold, insightful in many ways, interested in many things, and endlessly skeptical. "The most brilliant, the most secretive, the most discrete, the most sensitive, and the most rigorous of French psychiatrists," a colleague wrote after his death. Delay was a true "artist of medicine."

As a young man he had wanted to be a writer, and in addition to his work in mental health he would pen fourteen works of literature, including well-received novels and biographies, an effort that would win him election to that intellectual zenith of literature and thought, the French Academy.

And so Delay, a forceful figure in an elegant dark suit, watched over Sainte-Anne, evaluating the scene as if from a distance, weighing, analyzing, and transforming the hot simmer of patients into columns of cool facts, keeping his feelings to himself, focusing on research that could help, looking for measurable results.

Delay was careful, correct, and precise in everything. Freud and his followers might have made a fad of psychoanalysis and talk therapy, and wealthy neurotics might get some relief from talking about their dreams and sex lives, but Delay knew that this meant nothing in a mental asylum. His patients had deeper problems, likely rooted in physical dysfunctions of their brains. Delay believed that severe mental illness came from biology, not personal experience. He was, for his day, a revolutionary who wanted to free psychiatry from Freud's woolly thinking and unproven theories and move it toward becoming a real science, rooted in measurement and statistics, able to take its place proudly among the accepted fields of medicine. The keys, he believed, would be found in the tissues and chemicals of the brain.

But his brilliance and beliefs had yielded little in the way of cures. This was a failure rooted in the same problem that faced all psychiatrists: In the end, no one knew what caused madness. Finding cures, therefore, was a near impossibility. Psychiatrists ended up trying almost any therapy in hopes of finding something that worked, but nothing seemed able to alter the trajectory of deep madness. Many asylum doctors and staff members became despondent after years of butting their heads against the wall; depression was common among caregivers, and suicide was not unknown. It came from their inability to help those most in need of help. One of Delay's top lieutenants felt this way after a decade working at Sainte-Anne: "What I had learned in nearly ten years did not help me at all in treating mental illness. . . . I was a powerless onlooker."

Beautiful calm

The raving and thrashing young man had been in and out of Val-de-Grâce twice already, and both times the doctors at Laborit's Paris military hospital had done all they could for him: sedatives, anesthetic treatments, insulin coma treatments, and twenty-four electroshock sessions. "Jacques Lh," as he was called in their reports, would begin to respond, get a little calmer, and they would let him out. A few weeks later he would be back, out of control, threatening violence. So this time, in January 1950, they tried something new: Laborit's experimental drug, RP-4560. No one knew how much to give. For his surgical patients, Laborit had found that 5 to 10 milligrams worked well. So the Val-de-Grâce psychiatrists gave Jacques ten times that much. Within a few hours, the young man was asleep. And when he woke, to the physician's astonishment, he remained calm for eighteen hours before falling back into his mania. They gave him another dose of Laborit's drug, and another, as often as they thought necessary, at levels they hoped would work. They tossed in some sedatives and whatever else they thought might help. And something strange happened. Jacques's calm periods lengthened. By the end of three weeks, his condition had improved so dramatically that he was, as the reports noted, rational enough to play bridge. And so he was released.

When an account of this unusual single case involving an experimental drug was published later that year, it caused a minor stir in psychiatric circles. Some physicians were eager to test Laborit's drug. But others were deeply suspicious—both of the whole idea of drug therapy for mental disease (other than putting patients to sleep, there had been an unbroken history of failed drugs) and of Laborit himself. Laborit might be brilliant, but he was also seen as a little too sure of himself, a little too cocky. He had been publishing his successes with RP-4560 in surgery, pushing his artificial hibernation approach. He more than hinted that the drug might have applications in the mental health field. But Laborit was not a psychiatrist, had little training in mental health, and was far from an expert. To the psychiatric leaders in France, he was a surgeon with some odd ideas. What do surgeons know about the human mind?

Still, these were interesting results. RP-4560 trickled out into the medical community, eagerly shared with interested physicians by the drug's maker, Rhône-Poulenc. Through 1951, RP-4560 was tried on a number of patients with a variety of problems, and a surprising number of them seemed to get better. It helped relieve itching and anxiety in a patient with eczema. It helped stop the vomiting of a pregnant woman. And it seemed to work across a wide range of mental patients: It was tried on neurotics, psychotics, depressives, schizophrenics, catatonics—even patients thought to have psychosomatic disorders. Dosages were guess-and-test; durations of treatment were uncertain. Sometimes the drug did nothing. But many times it helped.

And in some cases, the effects seemed miraculous.

What was needed next were large-scale tests by reputable experts. It was the beginning of a year that one historian of psychology called "The French Revolution of 1952."

JEAN DELAY, like Laborit, was interested in the general idea of shock. But his interest focused on the beneficial mental effects of various types of shocks. Shock treatments were all the rage in mental asylums. In 1952 the focus was on electroshock (more properly, electroconvulsive therapy, or ECT). But there were other techniques that used drugs or even induced fevers to create a shock effect. In some cases, for no reasons anyone really understood, these treatments led to striking improvements. But only in some cases. Often shock seemed to do no good at all.

Delay wanted something better. He was an early proponent of ECT. He had seen some severely ill mental patients emerge from ECT sessions much improved and better able to function. But even under the most careful conditions, there were still many failures. And in the early days, ECT was close to barbaric and often dangerous. ECT patients jerked and writhed as the electricity jolted them. Some went into spasms so strong they broke bones. Some died.

Delay, always on the lookout for biological treatments, was also more willing than most psychiatrists to experiment with drugs. His staff tried various chemicals to treat depression and catatonia. Delay personally experimented with LSD soon after its discovery, and in the early 1950s his people tested the effects of mescaline on both normal and psychiatric patients. Drugs were useful tools, Delay believed.

Sainte-Anne was a good place to try new things. One day in late 1951, one of Delay's top men, Pierre Deniker, walked in with a story about his brother-in-law, a surgeon, who had heard about some new methods for preventing shock that were being tried out at the military hospital. The fellow doing this work, Laborit, reported patients who were calm and passive when cooled and given a cocktail of drugs. The brother-in-law told Deniker, "You can do what you like with them." And Deniker, like Laborit, thought about treating mental patients with the drug. Perhaps it could help calm the most agitated, confused, and violent among them. Sainte-Anne started testing Laborit's drug, RP-4560. The first patient was Giovanni A., a fifty-seven-year-old laborer brought in by the police in March 1952, raving and incoherent. He had been creating a disturbance in the streets and cafés of Paris, wearing a flowerpot on his head and shouting nonsense at people. He looked like a schizophrenic, an incurable.

Under Deniker's supervision, he was given a shot of RP-4560, laid down, and cooled with ice packs. Giovanni stopped shouting. He grew calm, seemed to fall into a daze, as if he were watching everything around him from a distance. He slept. The next day they repeated the procedure. He remained calm as long as he received regular doses of the drug. And he gradually got better. His fits of yelling and babbling became fewer. After nine days he was able to have a normal conversation with his doctors. After three weeks, he was discharged.

No one at Sainte-Anne had ever seen anything like it. It was as if Giovanni had regained his lost sanity—as if Giovanni the incurable had somehow been cured. Deniker quickly tried the new drug on more patients. At first he continued mimicking Laborit's artificial hibernation approach, cooling the patients with ice packs after their injection, using so much ice that pharmacy services had trouble keeping up. But his nurses, annoyed by the constant attention the ice required, suggested trying RP-4560 without the cooling. They found they didn't need the ice; with mental patients, the drug alone worked just as well.

The nurses loved RP-4560. One or two shots turned even the most difficult and dangerous patients into meek lambs. Deniker and Delay respected their nurses; they knew they were onto something special with this drug when the head nurse came to them, impressed, and asked, "What is this new drug?" You couldn't fool nurses.

Delay took a personal interest in the work and was often at Deniker's side. They expanded their tests. Every case was carefully tracked, the results meticulously charted.

And patterns began to emerge. Yes, RP-4560 helped patients sleep, but not like a standard sleeping pill. It didn't knock them out. It left them "steeped in sweet indifference," as Delay put it—conscious, better able to communicate, but distanced from their madness. With that distance often came the ability to reason: Over time, the drug made many patients less mentally confused and more coherent.

They started trying it on the most severe cases at Sainte-Anne, including the incurables, patients who had been locked away for years, suffering from deep depression, catatonia (in which patients stopped moving or responding), schizophrenia, and any sort of psychosis that was not responsive to other therapies. In every case, they noted, the drug had "a powerful and selective calming effect."

One major problem with the deeply mad was that doctors could simply not talk to them. Without that communication, therapies were limited. The real revolution began as many Sainte-Anne patients—not all, but many—began to talk with their doctors. Their wits had returned. RP-4560 did more than calm patients down. It "dissolved delirium and hallucinations," one physician marveled. "We were astonished and fired with enthusiasm by these results," remembered another.

Almost as important as the effect on patients was the effect on the staff. Asylum doctors and nurses, accustomed to constant noise in the back wards, punctuated by outbursts and screams, found themselves in a new world, a much quieter, calmer place where progress was possible. Accustomed to accepting that many of their patients would never be cured, they suddenly found themselves able to communicate, to move cases forward, to give patients hope.

The most touching incidents involved incurables who had been locked away for years, fated to die in the asylum. When they got their first shots of RP-4560 and began to regain their senses, it was as if Rip Van Winkle were waking up. When they were, for the first time in years, able to talk, and were asked, "What year is it?" they would answer with the long-ago date when they first came to Sainte-Anne. Now they rejoined the world, understood what had happened to them, began to communicate, to listen to something other than the voices in their heads, to take part in occupational therapy, to talk through their problems. They began to heal.

These effects were so stunning that no one outside of Sainte-Anne would have believed them if Delay hadn't announced they were real. Delay's intellectual eminence and reputation for careful research commanded attention. He delivered his first results with RP-4560 on a fine spring day in 1952, in the elegant mansion of the Académie Nationale de Chirurgie on the Rue de Seine. Curiosity was intense, and the audience included most of France's top psychiatrists and psychologists. Delay delivered a clear and elegant talk that astounded his listeners and ignited a firestorm of interest.

Somewhat oddly, although he credited the work of several other early researchers, Delay did not mention Laborit's name. Laborit and his colleagues at the military hospital were offended by the slight, which became the start of a small personal and professional battle for scientific credit that smoldered for years. The fact was that they both deserved credit: Laborit spurred the creation of RP-4560 and suggested its value; Delay's work legitimized it for psychiatric care and introduced it to the world.

Between May and October 1952, Delay and Deniker published six articles detailing their early tests on dozens of patients suffering from mania, acute psychosis, insomnia, depression, and agitation. A picture emerged: This was an important new advance for the treatment of some, but not all, mental disturbances. It was especially valuable in the treatment of mania, confusion, and possibly schizophrenia. But it did not work for depression. And, like all drugs, it had side effects: Too much given over time could leave patients too drowsy, too indifferent, too unemotional—it could make them into zombies.

More and more physicians began asking for samples of the experimental drug, and Rhône-Poulenc was happy to oblige. It was tested throughout France, then spread into the rest of Europe. Reports came back of a startlingly wide range of effects, many of them outside of psychiatry. As Laborit had found, it was good in prepping patients for surgery and seemed to boost the effects of anesthetics, making it possible to lower their doses. It was an aid in sleep therapy, eased motion sickness, helped calm the nausea and vomiting of expectant mothers. And everyone agreed that it appeared to be remarkably safe.

Rhône-Poulenc wasn't quite sure what to do with all this good news. RP-4560 did so many things that the company couldn't decide how to market it. So they put it on the market in the fall of 1952 under the vague heading of "a new nervous system modifier," a bit like a narcotic, a bit like a hypnotic, a sedative, a painkiller, an anti-vomiting drug, and a booster of anesthetics rolled up into one. All this, plus there were positive effects on mental illness. It was good for surgeons, obstetricians, and psychiatrists alike. What kind of trade name do you use for a drug like that? Something vague, something hinting at big things. So it was released as Megaphen in France and Largactil ("large action") in the United Kingdom. But most physicians called it by its new chemical name, chlorpromazine (CPZ).

Psychiatrists and other mental health workers had been waiting for decades for their miracle drug, something that would do for mental illness what antibiotics did for infections, antihistamines did for allergies, and synthetic insulin did for diabetes. CPZ looked like what they had been waiting for.

This all happened before adequate animal tests were done, before any knowledge of how CPZ worked in the body, and without knowing whether, in the long term, it would prove safe.

Exodus

Rhône-Poulenc sold the American rights to their new drug to Smith, Kline & French (SKF), an aggressive, up-and-coming drugmaker. SKF got it ready for testing by the FDA. "They were so smart," one researcher said of the company's work. SKF submitted it to the FDA for the treatment of nausea and vomiting, saying nothing about mental health. That made approval a slam dunk; the FDA gave it a thumbs-up within weeks of submission in the spring of 1954. Once it was FDA approved, and thus deemed safe, physicians were free to prescribe the drug for whatever they wanted (this practice of "off-label" prescribing would become an important part of marketing many other drugs). SKF trade-named it, somewhat vaguely, Thorazine. And they began pushing hard for its use in mental hospitals.

SKF's job now was to sell the new drug not to the public, but to America's doctors. They put everything they had into it, launching a marketing blitz that became something of a legend. They flew Delay and Deniker over from France to give talks; created a fifty-member task force that organized medical meetings, lobbied hospital administrators, and put on events for state legislatures highlighting the possible use of the drug in lowering the asylum load. They made sure every new journal article noting positive effects for CPZ got a wide reading, encouraged research, and even produced a TV show, *The March of Medicine*, in which the president of SKF himself talked about the new drug's effects.

Advertisement for Thorazine

Thorazine "took off like a shot," one of SKF's directors remembered. SKF's PR department went into overdrive, pushing out the word to newspapers and magazines. "Wonder Drug of 1954?" a *Time* magazine story asked. The enthusiasm was fueled by real-world experience. Stories flew from doctor to doctor. There was the mental patient who hadn't said a word for thirty years, and after two weeks on Thorazine told his caregivers that the last thing he remembered was going over the top of a trench in World War I. Then he asked his doctor, "When am I getting out of here?"

"That," his doctor said, "was an honest-to-God miracle."

There was the physician who read the journals, saw the drug work, then took out a second mortgage on his house and put all the money into SKF stock. It was a good investment: The new drug was a blockbuster. By 1955, Thorazine alone accounted for one-third of

SKF's sales; the company had to go on a hiring spree and put up new production facilities to keep up with demand.

It was just a taste of what was coming. In 1958, *Fortune* magazine ranked SKF the number two corporation in America for net after-tax profit on invested capital. Its revenues shot up more than sixfold between 1953 and 1970, with Thorazine bringing in the lion's share of the money. The company pumped a good part of that profit back into research, building a state-of-the-art lab to find more mind drugs. Other companies did the same.

And mind drugs were suddenly everywhere. The term "mind drugs" as used in this book does not encompass every substance that can affect your mood or mental state, a roster that could include everything from your morning coffee to your evening cocktail, along with just about every street drug you can buy. The new mind drugs, the ones that first showed up in the 1950s, are legal drugs developed by pharmaceutical companies specifically to relieve mental disorders.

CPZ was the first, in 1952, becoming the first of a family of drugs we now call "antipsychotics." It was followed in short order by Miltown, the first everyday tranquilizer for treating minor anxiety, in 1955. Miltown was found by accident, when a researcher looking for a preservative for penicillin noted that some of his test rats were looking very relaxed. It became a sensation in the United States, a "martini in a pill" that could take the edge off stress, and it was quickly picked up by Hollywood stars—within a few years Jerry Lewis was making jokes about Miltown when he hosted the Oscars—high-level executives, and suburban wives. Other "minor tranquilizers" like Librium and Valium soon followed, the start of a popular craze for the pills the Rolling Stones called "Mother's Little Helper."

Then a Swiss researcher, working in the early 1950s on a cure for tuberculosis, noticed that some of his ill, depressed TB patients were dancing in the halls after they took one of his experimental drugs. It was called iproniazid, and it became one of the first antidepressants, hitting the market in the late 1950s and opening the door for Prozac and a flood of other antidepressants through the 1980s and 1990s.

Suddenly psychiatrists, who just a few years earlier had no drugs to get at the worst symptoms of mental disorders, had several new families of drugs to choose from. A whole new area of research, psychopharmacology, arose. Pushed by the kind of aggressive marketing to physicians that SKF had perfected with Thorazine, these new drugs all had their moments in the sun as they went through their Seige cycles—tranquilizers became signature drugs of the 1960s and 1970s; antidepressants bloomed into blockbuster drugs in the 1980s and 1990s; and the growing family of antipsychotics, which today includes Seroquel, Abilify, and Zyprexa, rank today among America's top-selling drugs.

Why did all of these mind drugs suddenly appear in the 1950s? Perhaps it had something to do with society's need to deal with the pain and stress of World War II, or the desire to escape from the conformity of the Eisenhower era. Whatever the reasons, the new mind drugs changed American attitudes toward taking pills. Now pharmaceuticals were not something you took only to battle a serious health problem: Now they were something you took after work to chill out, or over time to change your ability to cope with everyday reality. The mind drugs of the 1950s set the stage for the next wave of recreational drugs in the 1960s, when popping more colorful, more mind-expanding hallucinogens became a craze. Mind drugs changed American culture.

AND THEY CERTAINLY revolutionized mental health care. SKF's PR blitz for Thorazine helped make the drug a huge hit in public mental hospitals. At first psychiatrists had been slow to accept it, believing that no pill could really solve mental problems, that the road to mental health had to run through Freud and talk therapy, not drugs. Many psychiatrists argued that Thorazine simply masked the underlying problems, it didn't fix them. A split began to rupture the mental health community, with psychotherapists—followers of Freud, often in private practice, dealing with one patient at a time, often well paid—on one side, and asylum doctors—often in a public hospital, less well paid, and dealing with scores or hundreds of patients—on the other. The Freudians were in charge of much of the professional infrastructure for psychiatry in the 1950s, and "I can tell you the pioneers in psychopharmacology were looked upon as quacks and frauds," one of the drug pioneers said. "I was accused of being no different than the guys who sold snake oil in the Wild West days." The idea that a pill could be used to treat an organ as complex, as mysterious, as finely tuned as the human brain was unbelievable. Those who promoted such unbelievable chemical cures seemed no better than the old patent drug salesmen hawking their wares at small-town medicine shows.

It was the asylum doctors who really appreciated what CPZ could do. It was a breakthrough drug, something truly new, something that offered hope. As deeply ill patients became able to talk for the first time since their disease started, they told their caregivers things like "I can cope with the voices better" and "it brings me back into focus." While they might still hear voices and suffer from delusions, these symptoms didn't bother them as much. They could now talk about what they were experiencing. They could function.

As CPZ use spread, the straitjackets went into cupboards. Patients who were unreachable began to open up. One physician remembered a catatonic patient, a man who had spent years silently twisted into a strange posture that resembled an owl, getting a regimen of the drug. After a few weeks he greeted his doctor normally, then asked for some billiard balls. When he got them, he began juggling.

"Look, you can't imagine," said another early adopter. "You know we saw the unthinkable—hallucinations, delusions eliminated by a pill! . . . It was so new and so wonderful." By 1958, some mental health hospitals were spending 5 percent of their budgets on CPZ.

And then came the exodus.

For two centuries, the number of patients in asylums had risen inexorably. But in the late 1950s, to almost everyone's surprise, for the first time in history, the numbers started going down.

The two reasons were drugs and politics. The drugs, of course, were CPZ and all the copycat antipsychotics that followed. With them, doctors could keep patients' symptoms under control enough to allow them to leave the hospitals and return to their families and communities. Many were able to hold down jobs. Unlike opiates or sleeping pills, the new drugs were just about impossible to overdose on. Nobody would want to anyway, because antipsychotics do not make you euphoric. They simply allowed patients to tamp down their symptoms enough to function. None have ever been drugs of abuse. So instead of being housed for years in an asylum, patients could now be diagnosed, treated, given a prescription, and released.

The politics came from state and county budget-makers, who had long worried about the mushrooming costs of public mental health facilities. Getting patients out of the asylums and mental hospitals was a win-win: Patients got to live their lives, and taxpayers got out of paying a tremendous bill. If the asylums shrank, so would the tax burden. Money would be freed for other programs. Some would go to community-based counseling, which would keep in touch with the newly released patients, make sure they kept up on their drugs, and (it was hoped) track their success integrating back into society. The rest could be used for other priorities, like education.

The era of community-based mental health care started, and the old mental hospitals emptied. Thousands of patients were released every year, many carrying a prescription for CPZ. In 1955, there were more than a half-million patients in U.S. state and county mental health hospitals. By 1971, the number had been cut almost in half. By 1988 it was down by more than two-thirds. The giant old asylums on their green grounds were torn down or turned into luxury hotels.

The first years of this shift were a very strange time. Physicians who thought they'd never be able to help schizophrenic patients were watching them walk back into lives outside. Schizophrenic patients who never imagined leaving the asylum suddenly found themselves trying to piece together lives that had been shattered years earlier.

It was rarely easy. Suddenly patients were released, one physician remembered, to find that their husbands and wives were married again, that they had no jobs, that their ability to cope, while improved, was not what it was before they went in. Everything depended on taking their meds; if they did not, increasing numbers ended up back on the streets. While many newly released patients managed to successfully integrate back into their homes and communities, others did not. The situation was made worse when government agencies didn't adequately fund much needed community mental health efforts.

The exodus grew after 1965, when new Medicare and Medicaid programs offered coverage for nursing home care but not for specialty psychiatric care in state mental hospitals. That meant that tens of thousands of elderly mentally ill patients, many with Alzheimer's, were moved out of mental hospitals and into nursing homes, with the cost of their care shifted from state to federal budgets. The use of antipsychotics in nursing homes shot up. So did Medicare costs.

The dream of integrating mental patients back into society began fraying at the edges. Increasing numbers of younger patients, especially those who found themselves unable to live with their families, ended up housed in jail. More than half of male prisoners today, according to one recent survey, have been diagnosed with mental illness, along with three-quarters of female prisoners. Mentally ill homeless people can be seen on the streets of every American city and many smaller towns.

We're still dealing with the fallout. The number of beds in public mental health hospitals—designed to be available to poor people—has declined dramatically. At the same time, the number of beds in private mental health facilities—for the wealthy—have shot up.

CPZ changed the very soul of mental health care. In 1945, about two-thirds of the patients at Houston's Menninger Clinic took part in psychoanalysis or psychotherapy. In 1969, only 23 percent did. In the 1950s, most American medical schools had a few part-time psychiatrists on their faculty, and those few were often thought of as something akin to woolly-headed witch doctors by the rest of the professors. Today every American medical school has a full department of psychiatry.

Not that many people see a psychiatrist anymore. You don't need to in order to get a prescription for a mind drug. In 1955, just about anyone who went to their local doctor with a serious mental problem was referred immediately to a psychiatrist (who would likely put them into analysis). Today, most general practitioners are willing and often able to diagnose the problem themselves and prescribe a pill. In the 1950s, schizophrenia was blamed on bad parenting, emotionally cold "refrigerator mothers," and the home environment. Today, it's seen as a biochemical dysfunction that has little to do with parenting. In 1955, people with minor anxiety, minor depression, standard-issue worries or behavioral problems, trouble paying attention, or any of a thousand other minor mental problems, were expected to work through their issues with the help of their families and friends. Today, most of them take drugs.

For good or ill, CPZ changed it all.

IN THE TEN YEARS after it first reached the market, CPZ was taken by fifty million patients. But today it's hardly used at all.

It's been superseded by new formulations that have taken over the market, an evolution fueled by CPZ's negative side. The more the older drug was used in the 1950s and 1960s, the more patients started turning up with strange side effects. There was the "purple people" issue, when the skin of high-dose patients turned a strange sort of violet-gray. Others got rashes or developed sun sensitivity. In some, blood pressure would drop precipitously. Others developed jaundice or blurred vision.

These were relatively minor. Side effects were expected with any new drug, and most of CPZ's could be fixed with proper dosing. But then came something more troubling. Physicians around the world found that some of their long-term patients, maybe one in seven, again mostly those on higher doses, were getting twitchy, their tongues poking out uncontrollably, lips smacking, hands shaking, faces twisted into grimaces. They couldn't seem to stop moving, shifting from foot to foot, rocking in place. They walked with a jerky gait. It looked to some physicians like symptoms of encephalitis or Parkinson's disease. The condition, named tardive dyskinesia, was very serious. Even when the doctors lowered their doses, the symptoms could persist for weeks or months. In some patients they didn't go away even when the drug was stopped entirely.

So major drug companies searched for the next big antipsychotic, something that could do what CPZ did, but with added benefits and fewer side effects. There were twenty on the market by

1972. But none of this first wave was more than marginally better than the drug used by Laborit and Delay.

IN THE 1960S Jean Delay was at the height of his career. His work with CPZ had changed the world of medicine, he was widely respected, and he was being showered with a growing list of honors.

Then, on May 10, 1968, it all came tumbling down. Paris's May Revolution brought thousands of student revolutionaries into the streets, and some of them decided to take over Delay's office at Sainte-Anne. The students believed that madness was not so much biological, as Delay thought, as it was a social construct used to enforce conformity. Delay symbolized the establishment, powers that used CPZ like a "chemical straitjacket" to control anybody they deemed undesirable. Delay was everything that was wrong with psychiatry and society. The students pushed into the great man's office and shouted their ideas at him, emptied his desk drawers, tossed his papers in the air, and refused to leave. They occupied Delay's rooms for a month. Rumors said that they stripped his diplomas and awards from the wall and sold them as war booty in the square of the Sorbonne (when in fact one of his daughters had gone to his office and talked a student guard into letting her take most of them home). When Delay tried to lecture, they sat in the hall, playing chess and making rude comments. It was a humiliating public repudiation of his life's work.

It broke him. Delay gave up his position and never went back.

Laborit, in his own way, flourished. He never got over his resentment of Delay's downplaying his work with CPZ, a grudge he held for the rest of his life. But he went on to earn many honors of his own—including the Lasker Award for medicine, second in prestige only to the Nobel Prize—and became something of an outspoken hero, his hair modishly long, his comments on psychiatry free-flowing, his Gallic good looks earning him a moment as a movie star when he played himself in Alain Resnais's 1980 film, *Mon Oncle d'Amérique*.

THE ANTIPSYCHOTICS did more than empty the asylums and change the practice of psychiatry. They opened the door to studies of the brain that are continuing to shake our ideas of who we are.

The big question through the 1950s was: How does CPZ do what it does? It took a decade of research and a major change in how we view brain function to find the answer.

Before CPZ, most researchers viewed the brain as an electrical system, like a very complex switchboard with messages flashing over the wires (nerves). Things went wrong when the wires got messed up. Treatments like ECT could reboot the system. Lobotomies could cut out a faulty section of wiring.

After CPZ, scientists realized that the brain was less like a switchboard and more like a chemical laboratory. The trick was to keep the molecules in the mind in proper balance. Mental illness was redefined as a "chemical imbalance" in the brain, with shortages or surpluses of one chemical or another. Mind drugs worked by restoring the chemical balance.

Many years of intensive research showed that CPZ altered the levels of a class of molecules called neurotransmitters, which are essential in moving impulses from one nerve cell to the next. By using drugs like CPZ as tools to study brain chemistry, researchers have now identified more than one hundred different neurotransmitters; CPZ affected levels of dopamine and several others. Researchers at other drug companies began finding more antipsychotics that affected different arrays of neurotransmitters to varying degrees.

In the late 1990s a new string of antipsychotics began to appear with trade names like Abilify, Seroquel, and Zyprexa. These "second generation" antipsychotics weren't all that different than the first ones, including CPZ, but they did offer a somewhat lower risk of tardive dyskinesia. They were very effectively marketed as a great breakthrough. And, because they were somewhat safer, more doctors felt comfortable giving them to more patients and often prescribed off-label for conditions for which they had never been FDA approved: PTSD in veterans, eating disorders in children, anxiety and agitation in the elderly. Nursing homes, prisons, and foster homes began using the drugs to keep their charges quiet and under control. By 2008 antipsychotics had grown from a specialty drug used almost exclusively by severely ill mental patients to the bestselling class of drugs in the world.

The more mind drugs like CPZ were studied, the more they helped open up the chemical mysteries of the brain. And the more we learn about the breathtakingly complex brains we carry around, the less we seem to know. The human brain is the one system in the body that makes the immune system look simple. We've barely begun the long journey to understanding consciousness.

Perhaps more important, from a cultural angle, is how these drugs have changed our sense of who we are and how we relate to medicine. If our moods, our emotions, our mental abilities are simply chemical in nature, well, then we can change all that with chemistry. With drugs. Our mental states are no longer who we are. They are symptoms that can be treated. If we're anxious, we can take a drug for that. If we're depressed, we can take another. Trouble concentrating? Another.

Of course, it's not that simple. But many people are acting as if it is.

THE GOLDEN AGE

"THE NEWLY QUALIFIED doctor setting up practice in the 1930s had a dozen or so proven remedies with which to treat the multiplicity of different diseases he encountered every day," writes medical historian James Le Fanu. "Thirty years later, when the same doctor would have been approaching retirement, those dozen remedies had grown to over two thousand."

Those thirty years, from roughly the mid-1930s to the mid-1960s, marked what drug historians call the "golden age" of pharmaceutical development. These were the years when many of today's giant drug companies blossomed, hiring battalions of chemists, toxicologists, and pharmacologists, building enormous, cutting-edge laboratories, and employing offices full of marketing experts and patent attorneys. From these fast-growing corporations flowed what seemed to be a never-ending flood of miracle cures: antibiotics, antipsychotics, antihistamines, anticoagulants, anti-epilepsy drugs, anti-cancer drugs, hormones, diuretics, sedatives, painkillers—the possibilities seemed endless.

Thanks to antibiotics and vaccines, medical scientists had con-quered many of the infectious diseases that had plagued humanity since the beginning of time and were working on the rest. Thanks to antipsychotics and the new research into neurotransmitters, they had opened entirely new fields of study and approaches to the prob-lem of mental health. Now they were readying themselves to march on the last great killers, heart disease and cancer.

But just then, at the height of their success, drugmakers began to worry. Many of the golden age breakthroughs had come more or less by accident, as when the failed antihistamine drug was used to prevent surgical shock and then unexpectedly led to the antipsy-chotics, or when a penicillin preservative was found to be a tran-quilizer. These lucky breaks—drug historians like to use the word *serendipity*—led to billions of dollars in income, and in following these leads drug companies had created hundreds of similar drugs, increasing profits. Then the companies plowed much of that profit back into research and development, with the idea that more directed, more informed research could lead to the next great breakthroughs. The lucky breaks of the past would give way to a more rational, more targeted kind of research based not on playing with chemicals and hoping for something good to result, but from a greatly expanded understanding of the body and how things went wrong when dis-eases happened. Find out what went wrong in the body, identify the processes involved at the molecular level, and then design drugs to fight it. This was going to be the approach that would open the next golden age, one, it seemed in the 1960s, that was just on the horizon.

And yet . . . . there were hints that things might not work out as hoped. Take antibiotics, for instance. All the wonder-working done by antibiotics seemed to be reaching a sort of natural limit. The bacteria that antibiotics worked against were relatively simple creatures. They had only so many places to attack: The cell wall (where penicillin worked), their food processing system (where sulfa drugs worked), and so forth. To make more antibiotics, more points of attack would have to be found. And there were not an infinite number. Even when one was found, bacteria had a maddening ability to find ways to fight off the antibiotics, leading to resistance. Was there going to be an end to antibiotics?

There was, it turned out. During the thirty years from sulfa through the late 1960s, twelve new *classes* of antibiotics reached the market, with each class containing a number of brand-name variations. In the fifty years since then, only two new classes have been added. And very little money is going into new antibiotic development. This seems tragic in light of the growing problem of antibiotic resistance, and it is—but there are also good reasons for it.

Partly it's because all the low-hanging fruit has been gathered—all the easy targets have been identified and worked with. And part of the reason is financial. Finding new antibiotics is expensive, and the payback is relatively small. A round of the proper antibiotic clears up the underlying bacterial disease in a few weeks, after which the patient doesn't need any more medicine. That means no more sales. And no sales means that drug companies have little incentive to find new antibiotics.

The same concept of a limited number of targets applies to the human body as well. We are far more complex than bacteria, of course, sometimes dauntingly so (as in the case of the brain and the immune system). But that complexity is not infinite. The more scientists learned about the molecular workings of the body, the more they could see that here, too, the number of targets for drugs had its limits. They might be far away from reaching those limits, but they were there nonetheless. And when all the targets for serious diseases had been identified and drugs developed to treat them, why would anyone need new drugs?

At the same time, the fast-rising costs of developing new drugs meant that the behemoth drug companies needed big-selling drugs more than ever. So a subtle shift started, a move toward drugs—like tranquilizers, for instance—that didn't save lives so much as make lives more comfortable. The next great era of development, the richest in pharmaceutical history, would focus less on the quantity of life and more on its quality.

CHAPTER 7

SEX, DRUGS, AND
MORE DRUGS

THERE ARE THOUSANDS of drugs out there, but only one is known universally as "the Pill." It's something of an oddball drug: It doesn't exactly ease symptoms, like painkillers do, or save many lives, like antibiotics do. Its development was rooted as much in social activism as it was in medical research, and its significance for health pales in comparison to its enormous cultural impact. The Pill revolutionized the world's sexual habits and mores, opened a vast new range of opportunities for women, and—in ways beyond those of almost any other drug—changed our world.

Before the Pill, the joys of sex were almost inevitably tied to conception. The creation of life was still viewed by many people as the province of God as much as doctors. This did not stop people through history from trying to break the link between having sex and having babies. In ancient China, women drank solutions of lead and mercury to try to prevent pregnancy. In Classical Greece, pome-granate seeds were used as contraceptives (with a tie to the goddess

Persephone, whose eating of a pomegranate seed while imprisoned in the Underworld obliged her to return there for six months every year, leading to the infertile months of winter). European women in the Middle Ages wore weasel testicles on their thighs, wreaths of herbs, amulets of cat bones; they tried brews and ointments laced with menstrual blood; they walked three times around a spot where a pregnant wolf had urinated—all in attempts to prevent conception. It wasn't just that pregnancy and childbirth were leading causes of injury and death for young women, or that pregnancy out of wedlock was a sin. Getting pregnant meant the end of independence, a curtailing of opportunity, and the start of a lifetime of domestic responsibility. Anything that could prevent it, no matter how hopeless, was worth a try.

Once scientists got involved, things didn't get much better. In the 1700s and 1800s, the biology of pregnancy—all the events that happened inside a woman's womb during the nine months between conception and delivery—was a black box, a near-total mystery. Pregnancy itself could be avoided through abstinence, of course. But apart from that, the only successes in preventing conception came from equipping men with early forms of condoms, unreliable prophylactics made with everything from pickled sheep intestines to linen sacks tied around the penis with colorful ribbons.

In 1898 Sigmund Freud wrote, "Theoretically, one of the greatest triumphs of mankind would be the elevation of procreation into a voluntary and deliberate act." He spoke for a growing number of experts who appreciated, at the turn of the twentieth century, important new reasons for birth control: the looming threat of mass famine due to overpopulation, a growing movement for women's equality, and a desire among many leaders to rationalize and tame what seemed to be ungovernable impulses with unwanted outcomes—sex included.

Among this last group were officials of the Rockefeller Foundation in the United States, which in the 1930s began to lavish part of its enormous financial resources on the new field of molecular biology. One reason this effort appealed to businesspeople and scientists alike was that it promised a better understanding of the relationship between biology and behavior. "Psychobiology" was one of their buzzwords.

There were plenty of reasons to make the investment. The years between the two World Wars were a time of social and political unrest, economic depression, and growing worries about the threat of Communism, urban crime, declining morals, and the fraying of social bonds. Rockefeller officers wanted to better understand the role biology played, to find the genetic roots of criminality and mental disease, to lay bare the links between molecules, actions, and emotions. There was more than pure science involved here; the powerful men who ran and advised the Foundation also wanted to use what they learned to create a more rational, less impulsive world, one less likely to fall apart—and, as a side benefit, one that would be more favorable for doing business. Some rather disquieting first steps into the world of biological social control were bundled by the Foundation in the late 1920s into a program they called "The Science of Man." As science historian Lily Kay writes, "The motivation behind the enormous investment in the [Rockefeller Foundation's] new agenda was to develop the human sciences as a comprehensive explanatory and applied framework of social control grounded in the natural, medical, and social sciences."

Among the many things the Foundation funded were investigations of the biology around sex. Sex hormones were just beginning to be understood. Everybody knew that at puberty, human bodies underwent major changes, growing hair in new places, becoming fertile, and developing a fascination with sex. Many of

165

these changes appeared to be moderated by molecules in the blood that carried messages from glands to other organ systems. Those molecules—hormones—started percolating during puberty, then went through an absolute riot in females during pregnancy. In the 1920s and 1930s, researchers were just beginning to understand why and how it all happened, and what the major players were.

One important clue came from Ludwig Haberlandt, a thin, intense, mustachioed Austrian physiologist who used Rockefeller Foundation funds to support his work on hormone research. It was well known in the 1920s, for instance, that once a female got pregnant, she couldn't get pregnant again until after she delivered. In scientific terms, she was temporarily sterile. While pregnant, females stop ovulating (releasing eggs to be fertilized). Haberlandt found that he could make this happen in the lab, without pregnancy, by transplanting into nonpregnant female test animals bits of the ovaries from pregnant ones. Those bits of tissue appeared to be releasing something, some sort of chemical messenger—Haberlandt thought it was probably a hormone—that prevented ovulation. He made female test animals temporarily sterile. And he knew what he was aiming toward: isolating that hormone, purifying it, and making it into a birth control pill.

But he was a man before his time. The relatively primitive lab setups and chemical technologies available in the late 1920s were not up to the task of studying biomolecules at the needed level of sophistication; this lack of good tools and the early stage of scientific research into the chemistry of pregnancy slowed his progress. But it didn't stop him from publishing his ideas. In 1931 he wrote a short book about his work that outlined, "in uncanny detail," according to one expert, "the contraceptive revolution of some thirty years later." Haberlandt is now often called "the grandfather of the Pill."

When he was alive, his work raised a firestorm of criticism in Austria. "Accused of a crime against the unborn life," his granddaughter wrote, "caught in the crossfire of the moral, ethical, ecclesiastical, and political ideas of the time," he became a target of those who believed that procreation was God's work, not something that humans should try to control. Just one year after his prescient book was published, Haberlandt committed suicide.

His work was carried on by others. Within a few years no fewer than four research groups had isolated the molecule he had been searching for, the hormone progesterone. Other research groups followed that lead, trying to understand how progesterone worked in the body. During the 1930s, scientists figured out how progesterone and other sex hormones like testosterone and estradiol are built. They were all related, part of the chemical family of steroids, and were all built from five- and six-sided rings of carbon with different side chains attached. Steroid chemists still call the 1930s "The Decade of the Sex Hormones." Then World War II shifted research priorities to military needs, funding declined, and research into sex hormones slowed. The emphasis immediately after the war was on having lots of children, not preventing them. One of the few scientists who kept working hard at the chemical angle on contraception was Gregory Pincus, who cofounded a private research group, the Worcester Foundation for Experimental Biology in Massachusetts, in 1944. Like Haberlandt, Pincus and his close colleague Min Chueh Chang, a Chinese immigrant, were fascinated by hormones that might interfere with ovulation.

In the early 1950s, their efforts were supercharged with a jolt of energy and money, thanks to the efforts of the famed social activist Margaret Sanger. This legendary figure had earned a worldwide reputation for her decades of work for women's rights, especially the right to vote and birth control. She had been arrested after opening America's first birth control clinic in 1916, fought her case through the courts, founded the organization that would evolve into Planned Parenthood, and rallied other women to her cause. Her work was helped by an old friend, Katharine McCormick, an equally committed women's rights activist and heir to the enormous International Harvester fortune. McCormick, one of the richest women in the world, dedicated a good chunk of her money to supporting Sanger's work.

Margaret Sanger. Bain News Service, 1916. Courtesy: Library of Congress

Sanger and McCormick, then in their seventies, got in touch with Gregory Pincus in 1951. The two women felt that the time was right for a final, all-out attempt to make a birth control medicine. Their motivations included a desire to end the scourge of back-alley abortions; a dedication to making birth control safe, dependable, and affordable; and a belief that women, not men, should decide when and whether to get pregnant.

It was not going to be easy. The United States was home to the Comstock Laws, a rambling package of anti-vice measures designed in 1873 to suppress obscene literature and "Articles of Immoral Use." The Comstock Laws had been used in 1917 to shut down Sanger's first family planning clinic in Brooklyn, only ten days after she

Katharine McCormick, Mrs. Stanley
McCormick. Bain News Service.
Courtesy: Library of Congress

opened it. Sanger and McCormick had been battling for decades against all the "Comstockery" that followed, a lawmaking passion at the state and local level for wiping out all forms of immoral and obscene behavior. Comstockery banned the sale of contraceptives in twenty-two states. Comstockery made it illegal in thirty states to run advertisements about birth control. In Massachusetts, where Pincus was doing his research, Comstockery meant that giving a single contraceptive pill to a woman could result in a $1,000 fine or five years in prison. And Comstockery meant you couldn't perform human tests of birth control in the United States.

Sanger and McCormick took on the whole thing. They would fight the laws if necessary, and look for alternative paths if needed. And they would fund the science they needed to make birth control work. After some discussion with Pincus about the possibilities of chemical controls for pregnancy, Sanger backed his research and

McCormick started supporting his work at the Worcester Foundation. The new money moved Pincus's work forward at a more rapid pace. He got together with John Rock, a gynecologist and fellow sex-hormone researcher, and began to zero in on progesterone as the way to make a birth control pill.

There were problems from the start. One was that progesterone, made in small amounts in the ovaries of animals, was hard to harvest and hard to purify. A lot of cows, sheep, and other animals had to be sacrificed to yield a little bit of the hormone, which made pure progesterone very expensive—more costly, gram for gram, than gold.

A second problem was that progesterone did not move out of the stomach and into the bloodstream very efficiently. Hardly any of it was absorbed into the body when given by mouth. That meant pills were a problem. If they wanted to use progesterone in a birth control pill, they would have to find some kind of chemical substitute.

The answer to the first issue—the scarcity and cost of progesterone—came from Mexico, where a little start-up drug company called Syntex found a way to purify steroids from a local strain of giant yam. Syntex had been started in 1944 by a pioneering, imaginative (one colleague called him "gutsy") American steroid chemist, Russell Marker, who was working on ways to turn plant steroids (plants make steroids, too—but they have to be chemically altered to be active in humans) into more valuable products. He scoured the world to find plants that would make large amounts of the starting material he needed. In late 1941, he found what he was looking for in a botany textbook, in the form of a weird plant found near a certain stream in Mexico. A picture with the text showed a root that bulged above the ground. The natives called it *cabeza de negro*, a kind of Mexican yam with a tuber as big as a person's head—or bigger. A single root could weigh more than two hundred pounds. Marker got himself to Mexico City, grabbed a series of crowded, clattering local buses to the town of Córdoba, and on the way crossed the stream he'd read about. Next to the stream was a country store. Marker convinced the owner to help him find samples of *cabeza de negro*.

He found the root, but the rest was a bit of a fiasco: He had no plant-gathering permit; after he gathered some root anyway, his samples were stolen; then he had to bribe a local policeman to get back just one of the roots—a big one that weighed fifty pounds. He smuggled it into the United States and started tests. It made good quantities of the starting material he needed. He figured out a novel way to turn that starting material into progesterone. And he started looking for a big drug company to back his scheme for making progesterone and other steroids out of *cabeza de negro*.

Nobody bit. So Marker and some partners started their own drug company, Syntex, in Mexico. He got the store owner by the stream to collect and dry about ten tons of the root. He arranged lab help to extract the material he wanted. And he ended up with more than six pounds of progesterone—the most ever produced to that time—a small fortune in hormone.

With an abundance of progesterone available, the door was open for accelerated research.

The next step was getting the ovulation-blocking hormone into the bloodstream. Syntex scientists began experimenting, creating new synthetic versions of progesterone. One, called progestin, acted like progesterone in preventing ovulation and—this was the important thing—progestin made it out of the stomach, making it highly active when given orally.

That was almost the last piece of the puzzle. But not quite. Animal studies showed that progestin, while effective, was also potentially dangerous because it sometimes triggered abnormal uterine bleeding. The solution came from another of those accidents that seem to happen time and again in drug research when researchers notice something puzzling in their research, then try to figure it out. The paradox was this: When they were purifying these progestin-like hormones, they found that the more pure their preparations were—the more carefully they ensured that all contaminating substances were filtered out—the worse the bleeding got. That made no sense at all, unless perhaps there was a contaminant that was inhibiting the bleeding. So they went back and studied the older, less pure preparations and found they were laced with a small amount of a form of another hormone, estrogen. Further tests proved the point: Giving a little bit of an estrogen-like molecule along with progestin helped control bleeding. That became part of the recipe for the Pill.

Putting all this information together, Pincus and the other Sanger-backed researchers at Worcester thought they finally had it: a birth control pill that got through the gut and delivered its medicine to the bloodstream, made mostly of a version of progestin, with a little dash of a synthetic variation of estrogen to prevent bleeding. It was time for clinical tests in women.

The final challenge was legal. They couldn't test their experimental birth control pill on women in the United States because of laws that made dispensing contraception illegal. If Pincus and Rock wanted to test on humans, they would have to go where Comstock Laws did not rule. They turned to Puerto Rico, which offered, as one historian put it, a "perfect mix of overpopulation and no prohibitive laws." There, in the spring of 1956, in the Río Piedras housing project, the first experimental version of the Pill was distributed to hundreds of women.

The Puerto Rico trials became something of a scandal. Women were given the medicine without adequate information about possible side effects (because little was known about them) and without any real chance to give informed consent. After the tests started, when women started coming forward to report headaches, nausea, dizziness, and blood clots, many of their personal stories were dismissed as coming from "unreliable historians." Pincus himself brushed aside many reports of minor side effects as the result of "hypochondria." But the side effects were real. One Puerto Rican woman died of heart failure during the trials.

To Pincus and other researchers, the question of informed consent was less important than the fact that the Pill worked brilliantly. The FDA quickly approved Enovid (the trade name of this early formulation) in 1957, but not for preventing pregnancy. In order to avoid Comstock trouble, the idea of preventing pregnancy was avoided or considered as a side effect. The official approval for the drug came for the purpose of regulating menstruation—a classification that was accurate and avoided the mention of birth control, thus making the drug available in Comstock states. By 1960, when the FDA finally gave its official approval for the Pill's use in birth control, hundreds of thousands of women were already taking it. And after full approval, the drug really took off. By 1967, thirteen million women around the world had taken some form of the Pill. The number of users today, with greatly improved formulations, tops one hundred million.

Today's versions of the Pill arose, in part, to address the troubling side effect of heart problems in young women, including a significant increase in the risk of heart attack. While the total number of women suffering serious heart problems is still relatively small—mostly because heart attacks are rare in young women to begin with—the increased risk was very real. Issues with blood clots and heart problems led Norway and the USSR to ban the sale of the Pill in 1962. The problem, although less severe in more recent formulations, still exists, although no one is quite sure why. As one expert recently wrote, "The debate on the precise effects of different hormonal contraceptives on the haemostatic system is still going on."

Regardless of side effects, use of the Pill skyrocketed, and profound cultural effects followed. As hoped, the drug decoupled having intercourse from having babies. "The Pill enabled young men and women to put off marriage while not having to put off sex," as one recent journal article noted. "Sex no longer had to be packaged with commitment devices," like marriage rings. It was the beginning of the Sexual Revolution.

At a deeper level, the Pill opened up new opportunities for women. Once women gained the power to control pregnancy, they began to arrange different sorts of lives. One study found that after the Pill reached wide use in the 1970s, the number of women pursuing graduate degrees and professional careers shot up dramatically. The proportion of female lawyers and judges, for instance, rose from 5 percent in 1970 to almost 30 percent in 2000. Just over 9 percent of physicians were women in 1970; the number was almost 30 percent in 2000. The same pattern held for dentists, architects, engineers, and economists.

The Pill did not do all of that by itself, but it played an important role. Before its appearance, the old model for American women was to graduate high school and either marry immediately or put it off just a few years, perhaps long enough to get an undergraduate degree. A 2002 review by economists Claudia Goldin and Lawrence Katz found that after the advent of the Pill, the age of first marriage for women began to go up, just as women's participation in graduate programs began to do the same.

In a way, this completes a link between the 1920s men of the Rockefeller Foundation, with their aim of using biology as a tool for dealing with society's discontents, and the women's rights activism of Margaret Sanger and Katharine McCormick. Both groups wanted to use our growing scientific knowledge of the body and effects of drugs in order to achieve a social end. The difference was that the women wanted freedom and choice, while the men wanted control over unruly human impulses. The Pill offered women a way to get what they wanted. Now, thanks to a famous side effect, it was time for the men.

GILES BRINDLEY was one of those mildly kooky scientists, thin, balding, and bespectacled, an established researcher and the expert in the functioning of the eye, but also a composer of music and the inventor of an instrument that he called the "logical bassoon."

In addition, he had a strong interest in erections, which is how Brindley earned one of the odder footnotes in science history. It happened at a 1983 urology conference in Las Vegas, when he walked onstage dressed in a loose blue tracksuit, looked over the audience of about eighty people, and showed off his latest discovery.

His topic that day, he explained in his British accent, was erectile dysfunction, a big deal among urologists in the 1980s. Back then, nobody knew exactly how erections happened, or exactly what to do when they didn't. Nobody had a clear picture of what systems were interacting with what systems, or what chemicals were involved.

What people did know was that a lot of men had trouble with them, and that those troubles seemed to increase with age.

The only answers available at the time were mechanical: an array of pumps, balloons, plastic splints, and metal implants that were made to be inserted surgically, then pumped, folded, or snapped into position to create an artificial erection. Researchers were going to great lengths to find solutions that were comfortable for all parties involved. They were, for the most part, failing.

It might seem amusing now, but it was no laughing matter to the millions of men who suffered from some degree of erectile dysfunction. To them, it was a serious medical issue.

Enter Giles Brindley, polymath, logical bassoonist, and one of the last practitioners of the ancient and honorable tradition of medical self-experimentation. From Paracelsus and his laudanum to Albert Hofmann, the Swiss chemist who discovered LSD, doctors throughout history had often tried experimental drugs on themselves before involving innocent patients.

Brindley, in his fifties at the time, had been self-experimenting with his penis. Specifically, he had been injecting it with drugs in a search for something that would chemically, rather than mechanically, create an erection. And, he told his Las Vegas audience, he had been making progress. He showed thirty or so slides of the effects. Even at a urology gathering it seemed a bit adventurous (at least in the days before social media) to see a man so routinely sharing shots of his own member. But the audience took it in stride.

Until Brindley felt compelled to demonstrate his results. At the end of his slide set, he told the audience that just before coming down to the conference hall he had injected himself in his hotel room. He stepped around the podium and, to general dismay, pulled his running pants up tight to show the results.

"At this point," remembered one audience member, "I, and I believe everyone else in the room, was agog . . . I could scarcely believe what was happening on stage."

The good professor then looked down, shook his head, and said, "Unfortunately, this doesn't display the results clearly enough." And he dropped his pants.

There was not a sound in the room. "Everyone had stopped breathing," an attendee remembered. Brindley paused dramatically, then said, "I'd like to give some of the audience the opportunity to confirm the degree of tumescence." Pants around his knees, he shuffled off the stage and toward the audience. Some of the women in the front row threw up their arms and screamed.

Their cries seemed to awaken Brindley. Realizing the effect he was having, he hurriedly pulled up his pants, returned to the podium, and finished his lecture.

Brindley's idea of using a syringe to shoot drugs into the penis never caught on, and the plastic and metal contraptions being touted by other researchers survive, for the most part, only as medical curiosities. They have all been replaced by a new generation of drugs, led by a famous blue pill.

And it all happened—as is so often the case in drug discovery—by accident.

SANDWICH, A SMALL TOWN on England's south coast, is known mainly for its well-preserved medieval guildhall and a few nice tourist cafés. It is also home to a research center for Pfizer, one of the world's leading drug companies. There, in 1985, scientists were trying to figure out a new way to treat angina, the excruciating chest and arm pain caused by reduced blood flow from heart disease. The Sandwich team wanted to find a drug that could open up blood vessels so that blood could flow more easily, as a way to ease angina's pain.

It turned out to be a tough problem. Blood vessels react to a number of different chemicals in the body, with each chemical in turn tied to a cascade of reactions—one chemical that kicks off production of another that kicks off production of another and so on—and each cascade triggered by yet other chemical signals from other parts of the body. But, fearlessly, the Sandwichers of Pfizer forged on, focusing on reactions they knew were involved, finding others that were new to them, and searching for drugs that might relax the blood vessels around the heart without causing terrible side effects.

In 1988, after looking at thousands of candidate chemicals, they finally came up with one that looked pretty good. Substance UK-94280, which worked by blocking an enzyme that destroyed another chemical that was tied into the blood-vessel-relaxing business—all part of a terrifically complicated system—looked like it might be worth trying out on humans. So they tested it on coronary heart disease patients.

And, like most drugs in early development, it crashed and burned. As one researcher put it, the initial clinical performance "fell short of our expectations"—a nice way of saying that the experimental drug worked too erratically and had too many side effects. Higher doses gave patients everything from indigestion to unbearable headaches.

And there was another side effect related to blood flow that affected only the men in their test group: UK-94280 kicked off erections. A few days after a dose, male patients reported that while their heart symptoms might be unchanged, their sex lives were definitely better. "None of us at Pfizer thought much of this side effect at the time," one researcher recalled. "I remember thinking that even if it did work, who would want to take a drug on a Wednesday to get an erection on Saturday?"

Then somebody at Sandwich realized that opportunity was knocking. Executives at big drug companies like Pfizer were always on the lookout for the next big thing. It was a matter of creating the right drug at the right time for the market. Particular attention in the 1980s was being paid to the biggest potential market of all: aging baby boomers. Members of the post–World War II generation, the biggest population bulge in history, were in their forties now, looking ahead to their retirement years. When that happened, drugmakers wanted to be ready with a slate of new medicines for the ills of growing old.

Through the decade, research funds were poured into searches for anything that could treat the biggest problems of the elderly: heart disease, of course, but also arthritis, mental decline, kidney issues, baldness, wrinkles, cataracts, and so forth. The idea was not to find a chemical Fountain of Youth, some definitive cure for these conditions, so much as to treat the symptoms, ease the pain, lessen the severity, hold them at bay, make them bearable—improve the quality of life. Drugs like these would have the added benefit of longevity—not for the patients so much as for the prescriptions. Drugs to ease the symptoms of aging-related conditions would not be taken for a short time, like an antibiotic, but indefinitely, like a vitamin pill. Profits would roll in for decades. These "quality-of-life drugs" were where the big money was going to be. One of the big issues of later middle age was erectile dysfunction. Sixty percent of men in their sixties had at least some trouble getting it up some of the time, and the percentage rose with age. This was a huge potential market. Then along came UK-94280 and its unexpected side effect. Pfizer decided to keep working with the drug. Only now their interest was not angina.

How do you test a drug like that for effectiveness? Here's one way: Gather a group of men who suffer from erectile dysfunction (ED), strap some gear around their penises to measure girth and hardness, give them various doses of UK-94280, and let them watch porn. The results, in clinical language, were "encouraging."

Then there was Pfizer researcher Chris Wayman, who built a "model man" in his Sandwich lab, with electrical switches substituting for nerves and, in place of the privates, pieces of penile tissue taken from impotent men. Each swatch of tissue was stretched between two little wire hangers attached to a measuring device, then suspended in a liquid bath. It was now possible to measure the tension and relaxation of the tissue. What Wayman was looking for was relaxation. Relaxed blood vessels can carry more blood, and so are more capable of engorging the penis.

When UK-94280 was added to the solution and the electricity switched on, blood vessels in the little pieces of tissue relaxed, just as they would need to do in an erection. "Now we were onto something which could only be described as special," Wayman told the BBC. Pfizer gave their new experimental drug the scientific name sildenafil and pushed its development to human tests.

Its effectiveness came as something of a surprise. Erections in men are not simple affairs. A hard penis arises out of the interplay of mind and body, a lot of blood flow, and a dizzying array of chemical reactions. Arousal itself appears paradoxical: Instead of switching on the penis, arousal damps down signals that keep blood flow to the penis at a minimum. Rather than pumping more blood, it's a bit more like opening the floodgates on a dam. But that's just the start. You also have to relax the blood vessels so they can fill and become rigid. The arousal process signals nerves in the blood vessels to start a chemical chain reaction; at the end of that chain is cGMP, a molecule the body produces to relax arterial smooth muscle and allow engorgement.

The system also has to be reversible, of course, or else the subject, once aroused, would walk around all day with a raging hard-on. Something has to put the process in reverse. The body does this by making an enzyme that breaks down cGMP; when the level gets low enough, the erection goes away.

And that, it turned out, is where sildenafil comes in. It blocks the enzyme that breaks down cGMP, allowing the levels of this critical chemical to stay high enough to keep an erection going. It works especially well among men whose ability to produce cGMP has been damaged, as in some heart patients. It doesn't kick off erections all by itself—you still need erotic stimulation to get the ball rolling—but it keeps them going once they start.

Just as Pfizer was gearing up sildenafil for release to the public, the National Institutes of Health handed them a big gift. In a 1992 conference (later buttressed by an influential study published in

1994), experts decided to expand the medical definition of erectile dysfunction. No longer would ED be the complete failure to achieve erection (the old idea of "impotence"). It henceforth would encompass any inability to achieve an erection adequate for "satisfactory sex performance." The details of what that meant were left up to individual doctors and their patients. With this more subjective, more expansive definition of what could be considered a diagnosable illness, the universe of men with ED suddenly got very much larger. The pre-1992 market of around ten million impotent men tripled overnight, and now included about one-quarter of all men over age sixty-five.

The timing couldn't have been better for Pfizer. They pumped tens of millions of dollars into accelerated sildenafil testing involving thousands of men. The results "exceeded our wildest expectations," one researcher said. The drug did what it needed to do and did it with remarkably few side effects. Now it needed a trade name that would push sales. The company riffled back through its files and came up with Viagra, a name that had been brainstormed some time before, then filed away waiting for the right drug. It was perfect, with its hints of both male strength (vigor) and unstoppable rushing waters (Niagara).

Pfizer patented its new drug in 1996 and got FDA approval in 1998. It was apparent from the start that the company had a winner. Their marketing department had a field day with the new drug. On May 4, 1998, *Time* magazine put Viagra on its cover, with art of an older guy (looking vaguely like the comic Rodney Dangerfield) clutching a naked blonde while popping Pfizer's distinctively four-sided blue pill. The headline on the cover sounded like something the marketing and ad teams could only dream about: "The Potency Pill: Yes, VIAGRA works! And the craze says a lot about men, women, and sex." Inside, the reporters asked, "Could

The distinctive Viagra tablet. Photo
by Tim Rickman

there be a product more tailored to the easy-solution-loving, sexu-
ally insecure American psyche than this one?" That's what they call
free advertising.

Fueled by enthusiastic and mildly titillating media stories, sales
swelled. The first day Viagra became available, one Atlanta urologist
wrote three hundred prescriptions for his patients. Some accom-
modating physicians sped the process up, doing quickie fifty-buck
telephone exams with unseen patients, then writing them a script.
Most health insurance companies started covering the costs. The
New York Times called it "the most successful introduction ever for a
drug in the United States." Pfizer stock shot up 60 percent.

And it just kept getting better. Two years after its introduction,
Viagra was available in more than one hundred countries; doctors
were writing around thirty thousand prescriptions per day; more
than 150 million pills had been sold worldwide, and Viagra was
bringing in around $2 billion per year in sales. The "little blue pill"
was now standard equipment for an older man's night out.

Other companies saw Pfizer's success and immediately jumped
into the game. Cialis and Levitra showed up in 2003, slightly different molecules that worked pretty much the same way, on the same
target, but varied somewhat in terms of side effects and timing. Cialis, for instance, lasts longer in the body, allowing men more than
a day of effectiveness versus Viagra's four hours or so.

But Viagra remained the king of ED medicines, changing sexual patterns among the elderly, spurring a million jokes—and raising some important issues. One centered on insurance coverage. Viagra was covered by most health insurance plans when it appeared (a fact not lost on women, whose birth control pills were, largely, not covered). Why should men's sexual health be more important than women's? In 2012 President Barack Obama's Secretary of Health and Human Services answered the question by ruling that most employers had to pay for women's contraception in their health plans under the Affordable Care Act. And some health plans have stopped paying for Viagra (although many still do).

Next question: Why wasn't there a Viagra for women, too—something women could take to enhance sexual pleasure? Drugmakers have spent millions looking for one, but no winner has come up yet. The issue for women is not erectile dysfunction, but more often a condition called female sexual interest/arousal disorder (FSIAD), which is less about blood flow than it is about desire. Many of the women who suffer from it (and up to a fifth of all women do) don't fantasize about or desire sex. Drug researchers figure it's tied to hormonal and neurotransmitter networks in the brain and are working on solutions that will likely be less like Viagra and more like an antidepressant.

These drugs raise longtime questions about the relationship between the mind and the body. Is sexual dysfunction in the body or in the mind? Male impotence—seen before 1990 as a difficult psychological problem rooted in parenting issues and childhood trauma—is now viewed, in many cases, as a simple problem of bio-hydraulics. It's more mechanical than psychological. The female sexual response looks like a more complex problem, with stronger links to the mind. You can draw your own conclusions. But when it comes to sex, it seems for the moment that men are easy and women are hard.

Through the early 2000s, Viagra continued to dominate the market. Men seemed to buy it no matter how much it cost; its per-pill price mushroomed from $7 when it was first introduced to close to $50 today. It was so popular and so expensive that a sophisticated black market sprang up, with dozens of underground pharmacies offering cut-rate blue pills without a prescription. One Pfizer study estimated that around 80 percent of websites claiming to sell Viagra were actually selling counterfeit drugs made at unlicensed factories. These phony pills contained, along with varying amounts of silden-afil, everything from talcum powder and detergent to rat poison and road paint. In 2016 authorities in Poland raided a suspected black market production site; behind a fake cupboard investigators found the entrance to hidden passageways and rooms containing more than a million dollars' worth of drug-making and packaging machinery, along with around 100,000 counterfeit blue pills. They shut it down, but others took its place. Fake Viagra is big business. Let buyers beware.

It took a decade for the initial Viagra craze to begin to ease. Many users found that, while the pill worked, it also caused head-aches, occasional priapism (erections that kept going hours longer than they needed to), and other minor side effects. Competing drugs became available. And the novelty was wearing off. Men were discovering that an instant erection was not necessarily the cure for all their sexual problems. Chemistry in a pill might be great for self-confidence, but it was no substitute for the chemistry in a relationship.

By 2010, almost half of all men who got a Viagra prescription weren't getting it refilled. Sales of ED drugs began leveling off that year; Viagra peaked in 2012 with sales just over $2 billion, then started dropping. The honeymoon was over. About the same time, its patent protection ended outside of the United States (and is slated to expire in the States in 2020). A standard patent for a new drug in the United States lasts twenty years from the time a company applies for it, although drugmakers are getting to be experts at finding ways to extend the patent period. Once it ends, however, the drug falls off a "patent cliff," as people in the industry call it, and other companies are free to produce the same drug. Generic versions appear, competition heats up, and prices fall. It can mean billions in lost income for the company that held the original patent on the drug.

THE RISE AND FALL of Viagra teaches some lessons. The first is that drug companies need big blockbusters like Viagra to survive. A successful new drug is a rare thing: Only a fraction of potential drugs that make it to human trials end up getting approved by the FDA, and only one in three of those that make it to market earn enough money to pay off their cost of development. The cost of development is the key: A new drug today takes one or two decades to move from discovery to market and soaks up, on average, more than half a billion dollars of investment to get all the way to the drugstore, a cost of development that has risen tenfold since the 1970s. (There is some debate about just how drug companies figure and report these costs, and whether the figures are as high as the companies claim they are. The numbers I'm using here are middle of the road.) No matter how you figure it, it's hellishly expensive to find a successful new drug. Drugmakers have to concentrate on a few potential big winners that might pay the costs of all the money losers. Viagra was that kind of winner. So was Pfizer's next big seller, the arthritis drug Celebrex—marketed again to aging baby boomers—which generated even bigger profits. Drug companies need blockbusters to keep profits up and shareholders happy.

The second lesson is that the best way to get a long-lasting blockbuster is to make sure that it doesn't cure anything. Neither of the two big sellers from Pfizer just mentioned cure the underlying condition. Both erectile dysfunction and joint disease are painful in different ways, but neither is life-threatening. Viagra and Celebrex treat symptoms, not diseases.

Quality-of-life drugs that treat symptoms can be prescribed endlessly; if a patient stops taking them, the symptoms return. So they make money endlessly. Given the high costs of drug development, it's easy to understand why drugmakers want that kind of payoff. The need for profit skews the kinds of drugs that are developed. It explains why drugmakers are putting very little effort into finding desperately needed new antibiotics and a lot of money into finding drugs that can treat the symptoms of aging.

It's not that Big Pharma isn't looking for drugs that will save the lives of patients. They are, especially in cancer treatment. But they need blockbuster quality-of-life drugs like Viagra to fund the process.

And after all, saving lives isn't everything. "More than any pill ever to be dispensed, Viagra has played to the yearnings of American culture: eternal youth, sexual prowess, not to mention the longing for an easy fix," one essayist opined. "It is the perfect drug for our time."

CHAPTER 8

THE ENCHANTED RING

BIG PHARMA'S SEARCH for the holy grail of pain control—a drug with all the power of opiates but none of the addiction—has led us not to perfect pain control, but to the some of the highest levels of addiction and the worst epidemic of overdoses in American history.

The difference is that we've now moved from natural opiates—based on the sap of the poppy—to entirely new, totally synthetic substances, made to order in laboratories. These newer drugs (which fall under the larger heading of opioids rather than poppy-based opiates) are far more powerful and potentially more addictive than any of the opiates used by our great-grandparents. Designed in part to help cure opiate addiction, they've only made the problem worse.

THE FIRST was discovered, once again, in Germany, at the Hoechst labs in the late 1930s, just before World War II. The company wasn't looking for it. It was found, once again, by accident. And the reason was the tail of a mouse.

Instead of a painkiller, the Hoechst chemists were looking for a drug to ease muscle spasms. Their starting point was a family of molecules completely unlike opium. The chemists were deep into the usual grind, starting with a candidate molecule, then making variation after variation, testing each one on mice to see what happened. That was when one sharp-eyed researcher noticed something odd: The mice that got one of these experimental drugs were raising their tails in an S shape. Most scientists would have ignored it. But this particular researcher had done work with opium-related drugs, and he knew what mice did when they were high on opiates. They raised their tails in an S shape. If he hadn't known better, he would have said this new drug was morphine.

So the Hoechst team did more tests. And it quickly became clear that they had discovered something entirely new: a powerful painkiller that did not resemble, in its molecular structure, morphine or codeine or any other alkaloid. True, this new drug wasn't as strong as morphine, but it did provide significant pain relief. Instead of putting test animals into the usual opiate dreamy state, it seemed to jack them up, like cocaine. Most important—and here the Hoechst researchers probably crossed their fingers—early tests hinted that it might be far less addictive than morphine.

Maybe they had stumbled across the holy grail. They named it pethidine (in the United States, it's better known as meperidine), did some quick tests on humans, deemed it good, and put it on the market in Germany. The ads said it was a powerful painkiller with fewer side effects than morphine and no risk of addiction.

Wrong, it turned out, on both counts. Pethidine—sold after the war under the trade name Demerol—has a slew of side effects, can be dangerous because of drug–drug interactions, and is anything but nonaddictive. It was attractive as a drug of abuse because it not only killed pain, it made users feel energized. Because of the combination of side effects and abuse potential—plus the appearance of newer painkillers—pethidine doesn't get used much anymore.

But it opened the door to something new: the promise of molecules completely unlike morphine or heroin that just might, with a bit more work, be made nonaddictive. This had what one historian called "a tremendously stimulating effect on drug research."

The years around World War II were a great time to be in the pharmaceutical business. New drugs were arriving at a record pace. There were a number of reasons for the flowering of big drug companies just after the war. The government had put a lot of money into medical research during the war, looking for better ways to treat wounds and prevent disease among soldiers, to understand how high altitudes affected fliers and high pressures affected submarine crews, how oxygen levels could be more accurately measured and blood plasma might be made in a laboratory. All that money helped scientists develop new tools and improved methods for testing and analyzing the human body. The victory over Germany yielded more riches for research, opening up laboratories, unveiling patents, and bringing German scientists to the United States. The post-war economic boom helped fund an enormous expansion of scientific research at universities and public laboratories, which in turn fueled further improvements in chemistry. Freed from wartime priorities and handsomely funded, drug science leaped forward.

Much of the excitement in medical research centered around molecular biology, the new ability to study life in finer and finer detail down to the level of individual molecules involved in digestion, say, or hormonal processes, or nerve conduction. This shift downward in focus, deeper and deeper into the workings of individual cells, was capped, in a way, in 1953, when the unlikely trio of a gawky American grad student named James Watson, a talkative young British researcher named Francis Crick, and research done by a talented female scientist named Rosalind Franklin, uncovered the molecular structure of DNA, opening up a new era of genetics research.

The more that was known about the molecules of life, the more opportunities came up for finding drugs that might have an effect. This built a sense of optimism that there might be a drug for every disease. All we had to do was understand the diseases well enough at the level of molecules, and then we could make the right drugs to treat them.

So first, there were powerful new tools; second, there was a growing understanding of the molecules of life; and third, there was a lot of money. With every successful new drug came another infusion of cash into the industry. Drug companies were growing fast. This private-sector growth was complemented after the war by a massive influx of funding in the United States from the federal government, which started funneling tens of millions of dollars into basic medical research through the new National Institutes of Health. Those drug companies that best understood the new dynamic—were most up-to-date on the latest findings, had the best lobbyists, and were most innovative in their in-house research—prospered. Smaller firms, those without the resources to compete, went belly up or were bought up.

Hoechst prospered. After pethidine, the firm made more variations of their synthetic painkiller through the war years. And after hundreds of failures, they finally found another effective painkiller—five times better than pethidine—that looked like it might be nonaddictive. They named it amidon. But this new drug, too, had downsides, particularly its tendency to cause nausea. It never got much use.

Until after World War II, when amidon made it to the United States and became better known under a new name: methadone.

It was a somewhat unusual opioid: a decent, but not great, painkiller; capable of being taken by mouth; slow to act, taking some time to get up to speed in the body; and less euphoria-inducing than other forms. Plus it made a lot of patients feel nauseated. Early U.S. tests seemed to confirm the German findings that it was nonaddictive. But as it went into wider use, it became clear that methadone patients, just like morphine patients, needed larger and larger doses to get relief, and many of them formed a dependence. In 1947 it was put on the U.S. list of controlled drugs.

Methadone never made much money as a painkiller. But there was something else: Because it was more unpleasant than euphoric, and because it could be taken without a syringe, physicians began playing around with the idea of using methadone as a way to get addicts off of heroin. Addicts didn't like it much, but it did soothe some of the itch of withdrawal. By 1950, a few hospitals had started using it to treat heroin addiction.

Heroin had disappeared from American streets during WWII because opium supply lines were interrupted. The number of addicts in the United States dropped by 90 percent, from 200,000 before the war to around 20,000 in 1945. As *Time* magazine put it, "The war was probably the best thing that ever happened to drug addicts."

But once the war was over, supply lines to Asia were reestablished (the most famous one running from Turkey through France to the United States—the "French Connection") and heroin came back with a vengeance. In the 1950s, the drug moved from inner-city black neighborhoods to rich white suburbs, from jazz clubs to pool parties. Heroin was cool, hip, dangerous. And profitable. "Junk is the ideal product," William S. Burroughs wrote in 1959, "the ultimate merchandise. No sales talk necessary. The client will crawl through a sewer and beg to buy."

The bigger and whiter the heroin problem grew, the more the government became concerned. Tough-on-drugs types argued that the answers were harsher laws, zero tolerance, and more jail time, while many physicians and community activists made the case for detox and compassionate care. A 1963 President's Advisory Commission on Narcotics and Drug Abuse split the difference, recommending both more treatment for addicts and tougher sentences for dealers. The emphasis was on getting junkies off the street and off the drug, into prison or in detox. Once they were clean, the thinking went, they could stay off the drug.

Except they didn't. About three-quarters of heroin addicts, once they were out of detox and had access to the drug, relapsed within a few months. Serious heroin addiction is really, really hard to kick.

Then the drug-friendly sixties hit, and everything got worse. Between 1960 and 1970, the number of junkies in the United States shot from 50,000 to around 500,000.

That was when methadone made a comeback. While many doctors shied away from addiction treatment during the 1950s—remembering perhaps how doctors after the Harrison Act were jailed for prescribing morphine to treat addicts—a few were still treating addiction as a medical problem. U.S. Public Health Service hospitals, for instance, were steadfast in providing treatment. And it was there that an increasing number of doctors first tried methadone.

Replacing heroin with methadone offered several advantages: The synthetic drug lasted longer than morphine, so instead of shooting up four times a day, addicts got a single dose; there was no need for needles; and it often eased the physical craving for opiates without giving addicts the euphoric kick of heroin.

In 1963 a tough, thickset New York physician, Vincent Dole, won a grant to study drug treatments to fight heroin addiction. Just getting the grant wasn't easy at the time because the drugs he wanted to study—morphine and methadone—were controlled. As a Federal Bureau of Narcotics agent told him, Dole was breaking drug laws just doing these studies, and if he persisted, they'd probably have to put him out of business. Dole didn't back down, inviting the Feds to try and shut him down so he could sue them and get a proper court ruling.

Dole, his wife, the psychiatrist Marie Nyswander, and a newly minted young physician, Mary Jeanne Kreek, started their work. They quickly found that morphine didn't work as a heroin substitute; the addicts simply wanted more morphine. That wasn't true with methadone. First, the researchers could get patients on an effective dose, one that eased withdrawal and the craving for heroin, and then could keep them there. The addicts weren't asking for more. And second, their methadone patients, unlike those on morphine, were not nodding out or sitting passively while they waited for their next dose. They were active and engaged. They might even be able to get a job.

Dole's team tried slowly reducing the dose of methadone, seeing if they could wean their patients off the drug, get them completely clean. But it didn't work. They could get the dose down to a certain point, but no lower. Hit that critical amount, and withdrawal symptoms would start.

The answer was to maintain patients on methadone for years—maybe for the rest of their lives. It was a trade-off, one drug for another. And methadone was the better choice. On methadone, addicts weren't breaking the law to get money for a fix, weren't shooting up with dirty needles, and weren't overdosing. They could build a life.

In 1965, when Dole and Kreek first presented their results, heroin treatment entered a new era. The media picked up the story, inquiries from other doctors started coming in, and Methadone Maintenance Treatment (MMT) was touted as the answer to the heroin epidemic.

It was the Seige cycle all over again—wild enthusiasm followed by deep misgivings. Dole remembered the years from 1965 to 1970 as the honeymoon period. Physicians were clamoring to try MMT. Every big city wanted it. Not even the Bureau of Narcotics—which "carped, infiltrated, and attempted to discredit the program," Dole said—could stop the momentum.

Then MMT became a victim of its own popularity. In the early 1970s, methadone treatment spread so far so fast that it got out of control. It got picked up by overeager centers and sometimes unqualified practitioners to the point where, as Dole put it, "Things became disorderly." Too many programs were treating too many patients with too little oversight or discipline. In that atmosphere it quickly became clear that MMT was not a perfect answer. An anti-methadone reaction set in, not only from hardcore antidrug types, but from the addicts themselves. They didn't like the nausea. They didn't like the state control. Addicts even made up legends centering on the fact that methadone had been developed in Germany during the Nazi years; they nicknamed it "adolphine" and made up conspiracy theories about it. A lot of junkies refused to take methadone, and a lot of those who did relapsed, ending up back on heroin.

Then the sixties were over, and it was time to get tough on drugs again. Methadone treatment was put under increasing government oversight. Paperwork increased. Funding decreased. The emphasis shifted from indefinite maintenance to short-term control using methadone as a stepping-stone, a way to get addicts off their drug of choice and into other, perhaps curative, therapies: psychotherapy, behavioral therapy, twelve-step programs, prayer. The new goal was stopping drugs entirely, not doling them out for life. By the 1980s MMT was out of fashion. But more recently it's made a comeback. Worries about AIDS transmission led to an appreciation of its role in reducing the use of dirty needles. Funding loosened up again. A National Institutes of Health consensus report in 1997 outlined the proven benefits: less overall drug use; less criminal activity; fewer needle-associated diseases; and an increase in gainful employment. The NIH panel recommended that all opiate-dependent persons under legal jurisdiction should have access to MMT, and the treatment is now FDA approved and growing in use. As one expert notes, "Today the safety, effectiveness, and value of properly applied MMT is no more controversial than is the assertion that the Earth is round."

But nobody's arguing that it's perfect. Many addicts and their families still go into methadone treatment with the idea that they're going to be "cured," but more than half of methadone program graduates end up using opiates again after discharge, or go back into treatment to get more methadone—which, remember, is itself a synthetic opioid. Permanent success rates (if you define success as never taking an opioid again) hover around 10 percent or less.

And that is the hard reality for all of opium's children. Once addiction has started, it is punishingly difficult to stop. It's certainly true for heroin. And it's proving true for synthetic opioids as well.

DEMEROL AND METHADONE were just the start. In the 1950s, one of the great drug discoverers of all time set his mind to creating an even better painkiller. His name was Paul Janssen. And he succeeded so completely that his work is still rocking our society.

He was the son of a Belgian physician and had followed in his father's footsteps, graduating from medical school at the University of Ghent and planning to teach medicine. But he had a passion for chemistry and new ideas about drug development. So he gave up his teaching, borrowed money from his father, and started a small drug company.

Janssen, the man his friends called "Dr. Paul," was a rare talent. He had the heart of an old alchemist; his goal was always to strip molecules down to their smallest active component, get to the spirit of the molecule, then build something around this purified essence, adding to it in order to create ever-better variations. Janssen was a deep thinker, able to concentrate intensely, to focus his mind on a given problem without giving up until it was solved. But he was more than a lab rat. He was also a tough-minded businessman, a builder of companies, a man who linked the creativity of an artist/chemist with the money-minded care of an executive.

He noticed, for instance, that when you compared the molecular structure of natural opiates like morphine with the newer synthetics like pethidine, there was one bit that they shared, one structure within their structures that the two had in common. It was a six-sided ring of atoms called piperidine. Given the similarity of action between these two families of painkillers, he thought it was likely that this relatively simple structure—this "enchanted ring," as it came to be called—was the spirit of the opium-like drugs.

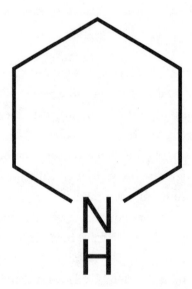

Piperidine, the "enchanted ring"

Janssen decided to improve it. He knew that older painkillers worked more slowly than they needed to and lost some effectiveness because they had trouble getting into the central nervous system. They were slowed because they couldn't easily get across cell membranes, which are mostly built of fat. So Janssen set out to make a fat-soluble opioid.

With that goal in mind, his lab began cranking out experimental drugs with the enchanted ring at the center, surrounded by side structures designed to be fat-soluble. They quickly found dozens of new drugs. In 1957, just after his thirtieth birthday, his fast-growing drug firm found a new opioid that was twenty-five times stronger than morphine and fifty times more powerful than Demerol, which worked more quickly and was cleared from the body faster. Phenoperidine, as his company named it, is still in use today as a general anesthetic.

And that was just the start. In 1960, Janssen's group synthesized another drug that was more than one hundred times as potent as morphine. It was, at the time of its discovery, the most powerful opioid in the world. This they named fentanyl, and they began working it into a whole family of new painkillers.

Janssen Pharmaceuticals discovered many other drugs as well—a groundbreaking new antipsychotic, anesthesia drugs, a diarrhea medicine used by astronauts in the Apollo program, antifungals, allergy medicines—finding more than eighty successful new drugs in all, four of which are on the World Health Organization's list of essential medicines. By the time Dr. Paul died in 2003, his company employed more than sixteen thousand workers around the world, and he had earned a reputation, as one of his colleagues wrote, as "the most prolific drug inventor of all time."

Janssen's company put fentanyl and its brother drugs into a variety of pills, skin patches, even lollipops for uses in controlling different amounts of pain in different patients. They remain standard medical tools for controlling pain. And they are all highly addictive, legally controlled drugs. As doctors and law enforcement have tightened legal access in recent years, fentanyl has gone underground, made in foreign countries and shipped into the United States. It's getting more and more common on the streets, in forms that can be snorted, swallowed, put onto blotter paper, used to spike heroin. Because it's so strong, overdoses keep going up along with increased use.

The spread of ever-more-powerful synthetics gave physicians better and better ways to control pain for their surgical patients, their cancer patients, and others with severe, intractable pain. And they also widened the door to more addiction in more people.

IF SCIENCE WASN'T going to solve the problem, then law enforcement would have to.

In 1971 President Richard Nixon announced his War on Drugs, including a large-scale offensive against opium products and traffickers. There was a mix of forces at play here: a backlash against the in-your-face drug use of the sixties; concern over the heroin addiction that veterans were bringing back from Vietnam; the rising appeal of law-and-order politics; and the growing realization that programs like methadone were having only limited success. His "Silent Majority" voting base, alarmed by what their children were getting into, by drug-related crime on the streets, and by drugs in the schools, wanted illegal drugs stamped out. There was a shift away from treating drug addiction like a disease. Increasingly, the public was liable to agree with writer Philip K. Dick, who wrote, "Drug misuse is not a disease, it is a decision, like the decision to step in front of a moving car. You would call that not a disease, but an error of judgment."

A decision, not a disease. From that perspective, Nixon's get-tough War on Drugs made sense.

It even gave the president a chance to show how "hip" he was by bringing celebrities like Elvis Presley into the White House to promote his move. Ironically, Elvis was taking a lot of drugs at the time. Nixon was gone soon afterward, but the Republican Party knew a winning political strategy when they saw it and made the War on Drugs a plank in their platform. "Just Say No," championed by Nancy Reagan, became the antidrug mantra of its time.

At the same time, a scientific breakthrough allowed scientists to finally figure out how opium works in the body. And with that knowledge came new hope for breaking addiction.

By the early 1970s, it was becoming clear that many processes in the body communicated with other processes, and that communication was done by molecules released by one cell and sensed by another. To deliver the message, specific molecules had to fit into specific receptors on the surface of cells. The old way of thinking was to imagine a key fitting into a lock. It's not quite like that in the body; it's maybe more like trying to fit different-shaped wooden pegs into different-shaped holes. You might not be able to put a big square peg into a round hole, but you could loosely fit a small square peg. Or you might be able to carve down a peg that's too big. In the body, the receptor system can be a little loose like that, recognizing and binding not just one perfect molecule, but also others that are similar. When the molecule binds to the receptor, it kicks off a reaction in the cell.

The great German physician/researcher Paul Ehrlich had theorized that communication happened this way in the body back in the late 1800s. But he and the next two generations of researchers had problems proving the point, because in the body many of the molecules that turn on receptors are made in very small quantities and are quickly broken down after they are made, disappearing to make room for the next set of reactions. This made them very difficult to study until the 1950s and 1960s, when far more sophisticated and sensitive lab equipment—X-ray and electron diffraction methods to study the structure of crystals; electron microscopes to study the architecture of cells; ultracentrifuges, electrophoresis setups, and chromatography equipment to separate molecules from one another; techniques for tagging molecules with radioactivity—made more sophisticated studies possible.

Including studies on opiates and other drugs. Many (but not all) drugs, it was found, did their work by activating receptors on the surface of cells. This was why certain drugs could have a specific effect on some cells, but not others. If a cell didn't have a receptor for the drug, nothing happened. If it did, reactions were triggered. Drugs could be used to find receptors and study them. They could also be tweaked, their structures changed slightly, to see what that did, allowing scientists to learn more about how drugs fit into receptors.

It was only logical to think that there must be receptors for morphine and other opium alkaloids. But it wasn't until 1973 that Solomon Snyder and a graduate student, Candace Pert, found them. Snyder was an MD with a strong interest in clinical psychology; he cut his teeth doing studies on LSD and other hallucinogens in the mid-1960s, trying, like everybody else, to figure out how such vanishingly small quantities of these drugs could produce such profound effects on the mind. He became an expert in doing experiments with molecules tagged with radioactive atoms. By following the radioactivity, he could follow the molecules in the body. He found that LSD, for instance, concentrated in certain parts of the brain after it was taken. Why did it go some places in the brain more than others? Turns out it was because that's where the receptors for LSD were. Snyder's lab at Johns Hopkins became a national leader in drug receptor studies.

Pert was a dynamic and determined young woman. Just before entering Johns Hopkins, she broke her back in a horseback-riding accident; her subsequent hospitalization gave her personal experience with the wonders of morphine. How did the drug do what it did? She maintained that interest when she started working in Snyder's lab as a grad student. As sometimes happens in science labs, friction developed between professor and student: Pert claimed that Snyder wanted her to work on insulin receptors and forbade her to work on morphine; she remembered being so fascinated with the substance that she worked on morphine receptors on her own initiative, even sneaking her five-year-old child into the lab so she could keep an eye on him while she worked at night. Snyder saw her as another grad student who should be doing whatever needed doing in his lab. In his memory, that included opioid studies. However, it worked—between them they found the receptor in the brain that fit opioids. And then they and other researchers found another. And another. The more they looked, the more opioid receptors there seemed to be—three major types have been found so far, plus several more variations (whether the total is three or nine is still being debated). Which raised a question: Why on earth have we evolved so many receptors in our brains for molecules that come from poppy plants? As Pert put it, "God presumably did not put an opiate receptor in our brains so that we could ultimately discover how to get high with opium."

It turns out that we didn't. In 1975 a pair of Scottish researchers found out that the brain itself made a natural chemical that those receptors were made to fit. It was called enkephalin, and it was only the first of a growing family of related molecules, all made in our own bodies, that we now call endorphins (for *endogenous morphine*). You can think of them as our body's own opiates. They play a vital role in helping us to control pain, calm down, and feel happy. They're the treat our body gives us when we do something nice for it: the molecules that make us feel good when we get a massage or have sex or experience a runner's high. They're even released when we laugh.

We make a bunch of them—different stimuli make them flow in different amounts at different times, and they react in varying ways with those different receptors. The result is a variety of effects that make it possible for our bodies to experience an exquisite array of natural pleasures.

Poppy alkaloids and the opiates we've made from them, plus the synthetics, all happen to trigger those same receptors. It's no wonder they're such bewitching drugs.

Snyder and Pert's early studies blossomed into entire fields of research. We now have far more refined tools for studying the receptors on our cells and the ways they can be stimulated or blocked. Much of modern drug-making is built around these studies. Existing drugs are often used to find the receptors; once found, the receptors can be studied to see what turns them on and off; the result is both new drugs and a better understanding of how the body works. It's a sort of virtuous cycle, with new drugs spurring a better understanding of the body, then that new understanding spurring a next round of better drugs. This is expensive, painstaking, and very important work. And it has led to hundreds of new medicines.

The discovery of opioid receptors and the molecules they work with also opened another door to pain control. Just as, seventy years earlier, organic chemists dreamed that some manipulation of the morphine structure might yield a nonaddictive substitute, now molecular biologists dreamed of another new path, one that led through the newly discovered opiate receptors. Receptors are turned on by molecules called "agonists"—morphine, heroin, oxycodone, and fentanyl are all agonists—but can also be turned off by "antagonists," molecules that attach to and block receptors without switching them on. When an antagonist blocks a receptor, it can't be turned on by anything else. Researchers found a way to do this with opioid receptors, developing antagonists like naloxone (marketed as Narcan). Naloxone attaches to opioid receptors but doesn't switch them on. One website compares taking Narcan to putting a piece of tape over the fingerprint scanner on your phone; you can put your finger on the scanner all you want, but the tape keeps the phone from getting the message.

Narcan attaches so strongly to opioid receptors that it can actually sort of muscle the real drugs off, taking their place, sticking tight, and preventing any more from switching on the receptor. That's why a dose of Narcan can save an addict's life. The opioid is still in excess in the bloodstream, looking for a receptor to land on, but can't find one. The resulting crash can be both horrible for drug users and near-miraculous for caregivers trying to save their lives; Narcan not only can wipe away all the euphoria from the opioid, throwing addicts into a form of instant withdrawal, but it can also stop an overdose in its tracks, pulling the victim back from the brink of death.

Researchers kept coming up with more and more new drugs that could modulate opiate receptors, new agonists and antagonists and partial agonists and agonist-antagonists (which have some properties of both), molecules specific for certain receptors and not others, molecules that acted different ways at different dosages, molecules that worked faster or slower, molecules that were quickly flushed out of the body and others that lasted a long time, a trove of new medicines that could turn receptors on and off selectively without using opiates.

In the 1970s and 1980s, the hope once again was that this fast-growing science might solve the whole heroin/opioid addiction problem.

But, no.

A RESPECTED EXPERT gives a speech at a medical gathering, reporting that America is the world's center of a growing drug crisis.

America consumes fifteen times as many opiates as Austria, Germany, and Italy combined; only 20 percent of these drugs are taken for legitimate medical reasons. There's evidence that almost a quarter of medical professionals themselves have some sort of personal opiate habit.

That's from a newspaper story that ran in 1913. Since then we've had more than a century of scientific research, social programs, and government pronouncements. And the problem has only gotten worse.

Today the United States, with less than 5 percent of the world's population, consumes 80 percent of the world's opioids. The number of prescriptions written for opioid drugs—both synthetic and non-synthetic—more than doubled between 1992 and 2015; the number of overdose deaths in the country went up almost fivefold over the same period. Today, opioid overdoses kill more Americans than car accidents and gun homicides put together.

How has this happened? Science plays a role. Drug companies keep looking for that magic combination of addiction-free painkilling, and they keep coming up short. As they explored, they found other things—more powerful, more targeted opioids—so the total number of available opioids and related drugs keeps going up year after year: specialized formulations that work fast or slow, pills that are time-released and pills that are coated to prevent abuse, pills that are tailored for all levels of pain. Trailing behind them are all the drugs that are not opioids but are designed to help treat opioid addiction (like methadone and buprenorphine); to reverse the action of opioids (like naloxone and others); to treat opioid-linked constipation; to energize opioid patients so they can get out of bed; to take the edge off the energized patients so they can get some sleep; and the list goes on.

Another major factor fueling the opioid epidemic is money. Prescription opioids are a $10 billion per year business; pain drugs in general in 2017 were second only to cancer drugs in terms of sales, with more than three hundred million prescriptions filled per year. That's not to mention the ancillary drug income, the illegal money from street drugs, the government program dollars, and money flowing through the burgeoning rehab, detox, and treatment businesses.

It's a huge industry. And most of the players have a vested interest in keeping their business going. So, as has been true for more than a century, drugmakers keep promoting the next anti-addiction tweak, rehab centers keep promising more effective programs, and the government keeps announcing new efforts to wage war on drugs. Most of these efforts seem eerily familiar to anyone who has studied the history of these drugs. President Donald Trump's recent idea about killing drug dealers, for instance, is the same one that was used by—and to some degree worked for—the Communists in China in the 1950s. Those kinds of programs are far easier to implement in centralized dictatorships than in Western democracies. Whatever the promised benefits of new formulations from drugmakers, newly redesigned drug rehab programs, or newly announced government initiatives, virtually none of these programs have, in any positive sense, worked. And the money keeps flowing.

If that sounds cynical, it is. Many, many people truly want to end this menace, and many organizations are honestly committed to bringing opioids under control and ending the scourge of addiction and overdose. But you can't get around the simple fact that money motivates many players.

And that includes doctors. Drug companies are masters at promoting their wares, and a lot of their effort goes toward convincing doctors to prescribe their latest and greatest. In the old days drugmakers would have ballyhooed their products, bought the doctor lunch, and offered a cigar. Today they offer the physician some payment as a consultant, or pay for some research; they invite the physician to a winter conference at a tropical resort, where other physicians—experts whose opinions support the drugmaker—highlight the results from supportive scientific studies. These studies might also be financially backed, the results sometimes tailored, and the final papers sometimes written with help from the drug companies. They make sure that the right information gets into the right journals. They might make sure that negative experimental results—the kind that might sink a promising drug—are smoothed over or buried. It's all very "scientific" and persuasive. And lucrative.

Physicians are also subject to trends in care. In the 1980s and 1990s, for instance, some leading experts in pain management argued that patients taking opioids for legitimate pain were unlikely to become addicted. The message of the day was: Prescribe until the pain is under control, even if the doses are high. Drugmakers obliged by coming up with more and more powerful variations on opioids, boosting the popularity of stronger semisynthetics like Oxycontin and synthetics like fentanyl. They became more and more common in medical practice.

Opioids were perfect for doctors who were increasingly short on time, especially the time needed for chronic pain patients, many of whom have complicated health histories and sometimes hard-to-diagnose reasons for their pain. Patients like that can eat up a lot of time talking about their condition; real answers can be very hard to find. A prescription for an opioid is an easy fix.

But far short of a perfect one. Patients would start on a relatively low dose, get relief, then find that they needed to increase their dose to get the same effect. They became drug tolerant. Their original pain was often replaced or increased by the pain of withdrawal, of simply not getting enough drug. In other words, it was easy for pain patients to become addicted.

But by the time that lesson became clear—and remember, this is the same lesson doctors grappled with in the 1840s with opium and the 1890s with morphine and the 1900s with legal heroin—in the first decade of the twenty-first century, opioid prescriptions were starting to go through the roof, followed by widespread dependence and addiction. The more that oxy and fentanyl were prescribed, the more they ended up on the street, either sold by patients with legitimate prescriptions or by dealers who found illegal ways to get them by the crate. Some addicts are experts at "doctor-shopping," taking their pain complaints to physician after physician, some of whom will run them out of their office, and some of whom will write them a script. Then the addicts take their duplicate prescriptions to multiple drugstores to get them filled. They take some and sell some. There's a huge black market for prescription opioids.

By 2010, the media and the public woke up to the fact that we were suffering through yet another opioid crisis. And the brakes were applied. In the past few years, consumption has dropped a bit. Physicians are prescribing less, moving away from the 1980s idea of "pain control no matter what" to a mind-set that more evenly matches risks with benefits. Government controls on the distribution of opioids have helped. Many drugmakers seem eager to cooperate with strategies to fight the epidemic and are seeking ways to curb abuse with better tracking of the flow of drugs from manufacturer to final user, and by continuing to make abuse-deterrent forms of opioids, with waxy coatings and time-release formulations that make it harder to get high.

But addicts, it turns out, are just as innovative as drug engineers. As soon as a new abuse-deterrent model of opioid comes out, somebody figures out how to smash, shave, microwave, snort, chew, or dissolve the drug to get past the deterrent and get their hit.

And that's the thing: The hit is always there. No matter how protected it is, at the heart of every opioid pain pill is the opioid itself. Taking the pill gets the drug sooner or later to the receptors in the brain. The drug attaches to the receptor, the receptor fires—and there's relief. The pain eases, the spirits soar, the jones is pushed away for a little while. There will always be some out there, on the street, as long as the poppy is harvested, the labs make synthetic versions, and doctors prescribe the drugs. And doctors will always prescribe the drugs because opioids are still, hands down, the best things we have to control pain.

In the end, if addicts can't get oxy or fentanyl or some other pharmaceutical-grade opioids, they can always fall back on heroin. Heroin use is exploding as the black market for prescription opioids is becoming more constrained. Many addicts, finding it harder to get a legal fix from their doctors after the recent round of watchdogging, simply switch to the old favorite. Today heroin is flooding the streets; it's cheap and available. Street price of a single pill of a strong opioid, Oxycontin or better, can run as high as $30 to $100 today. A bag of heroin, on the other hand, costs around $10, depending on your city. In many places, you can score a hit of heroin for less than a pack of cigarettes. And the heroin can be stronger than ever, boosted with a sprinkling of fentanyl or some other powerful synthetic. When you get it on the street, you never know how strong your fix is going to be. Overdoses have skyrocketed accordingly. The only winner seems to be the pharmaceutical industry. Every few years, drug companies come up with another opioid variation, some brand-new, can't-fail, anti-abuse version that promises a different result, just like heroin was going to fix the problem of morphine. As drug after drug fails, there's always another one to help switch addicts off the harder stuff, and untold millions of dollars are spent testing it out and trying to make some marginal difference.

Why America? Why are opioids America's particular problem, so much more than any other country's? Experts have been thinking about this question for decades, focusing on a few prime suspects. Part of the answer is rooted in the structure of our medical system, with its emphasis on short appointments for patients, reliance on powerful technologies, and bias toward finding a pill for every problem. Part of it comes from our economic system, with its insistence on increasing sales and profit. We are a wealthy society, and we can afford heavy pharmaceutical use. Part of it comes from our now-ingrained mind-set that drugs are a criminal problem, not a medical one. This funnels a lot of money toward the criminal justice system, police, DEA, and prisons, and dials back funding for medical approaches—clean-needle programs, addiction counseling, legalization of some drugs—that seem to work in other countries. There's also a bit related to our peculiar national character. We Americans love our freedom to do what we want, when we want, including taking the drugs we want.

And, disquietingly, there's the underlying fact that we're drawn to opioids for the same reason the Chinese were nearly two centuries ago: It's a way to escape. As one opioid expert put it, "We thought the big problem with these drugs is addiction. Now we realize the problem is with patients who take them and basically opt out of life."

And maybe it's because we're wimps. As a physician said during a recent symposium, "Americans think we should never be in pain." This is the flip side of our risk-taking adventurism. In part because of the quality of our drugs, we seem to have grown unaccustomed to pain and are unwilling to bear it. And not only physical pain. We are also lowering our tolerance for any sort of psychic discomfort, from minor anxiety to minor depression.

Increasingly, when we suffer any sort of discomfort, we pester our physicians for pills, and they prescribe them to us. This is not to say that millions of Americans do not suffer severe, long-term, very real pain, or severe depression, or crippling anxiety, and need opiates or antidepressants or tranquilizers to manage their illness. But in theory, a similar proportion of patients in every other culture or country should fall into the same category. The question is why American usage, both medically and on the street, is often so much higher. Are we in more pain than other nations? Are we suffering from more mental illness? There is little evidence that we are.

These issues are obviously complicated—as complicated as the workings of the human body—and dauntingly hard to tackle. Opioids are the ultimate case, because, as one expert concluded, "Opiate dependence is not a habit, nor is it a simple drive for some emotional craving. It is as fundamental to an addict's existence as food and water, a physiochemical fact: an addict's body is chemically reliant upon its drug, for opiates actually alter the body's chemistry so it cannot function properly without being periodically primed. A hunger for the drug forms when the quantity in the bloodstream falls below a certain level, the addict becoming anxious and irritable. Fail to feed the body and it deteriorates and may die from drug starvation." Read that again: Denied their fix, addicts are not just uncomfortable. They're starving.

Despite all the politicians' programs, the medical studies, the police task forces, and the social workers' best efforts, addiction rates have gone nowhere but up. The predictions are that Americans, as they age, will continue to take more, and stronger, opioids. Drug companies will continue to profit. And opium's thousand-year-old story will be rewritten for a new age.

CHAPTER 9

STATINS:
A PERSONAL STORY

IT LOOKED LIKE a piece of junk mail. Normally I'd just toss it, but the return address was from my local health-care system, so I ripped it open. Inside was a form letter from a doctor I'd never heard of. He was offering me a bit of unsolicited advice: Because my health records hinted that I might be at a higher-than-normal risk of heart disease, he wrote, I should consider taking a statin. He even included a helpful list of popular statins by name. He wasn't telling me what to do, but it was close.

Whoa. What? I was being advised by my health-care system to start taking a drug I knew nothing about in order to prevent a disease I didn't know I had? My personal doctor had never talked to me about statins during any of my annual exams. So why did I get that form letter?

My search for an answer to that question turned into a six-month odyssey, exploring a strange new area of today's big-money pharmaceuticals. It ended up teaching me about a major shift in how medicine is being practiced in America. It helped me understand more about today's prescription drug scene, gave me a few useful tricks to cut through the hype of drug ads, and drove home just how marginal the benefits of some highly recommended drug therapies can be. I came away surprised by some of the things I learned.

FIRST THINGS FIRST: Statins, it turns out, are amazing drugs. Their appearance in the 1980s represented a true breakthrough in medicine. They dramatically lower the amount of cholesterol in the blood and can help treat and prevent some of today's most devastating illnesses. They are taken by tens of millions of people around the world. They have received more study, on more patients, in more published papers, than almost any other class of drugs. They have saved tens of thousands of lives. Compared to most other prescription drugs, they have very mild side effects. And because many are off-patent and available as generics, they can be pretty cheap.

No wonder they've become huge international bestsellers. And yet . . .

As one leading heart doctor put it in a recent review of statins, "With over a million patient years of trial data and publications in the most prestigious medical journals, it is remarkable that quite so much debate remains as to their place in healthcare." The more data we gather, it seems, the less clear the conclusions.

That, and their enormous sales success, leads to some troubling questions. Are statins so great that, as some health experts advise, essentially everyone over the age of fifty-five should be taking one? They're relatively new—are there things we don't know about long-term side effects? Does taking a statin encourage people toward bad habits (as in, "I'm taking a statin, so I can eat whatever I want")?

And, at a more basic level, if lowering cholesterol is so good for you, why are experts still arguing about it?

The more I found out about statins, the more questions I had.

THE STATIN STORY started in the mid-1960s, when a Japanese college student named Akira Endo read a life-changing book, a biography of famed medical scientist Alexander Fleming, the man who discovered penicillin being released from a mold from the *Penicillium* family. What struck Endo was the idea that molds could make medicines. Molds, along with mushrooms, are kinds of fungus, and in Asia fungi had been used in health-giving foods and medicines for ages. What other important medicines might molds make?

Endo spent his life answering that question. When he was just starting on a career in drug research, he spent time at the Albert Einstein College of Medicine in New York City, where, in the cultural ferment of late-1960s America, he suffered a mild case of culture shock. Part of it came from the sheer wealth and power of the United States—the skyscrapers and hustle and money and music.

And part of if came from the food. "I was very surprised by the large number of elderly and overweight people and by the rather rich dietary habits of Americans compared to those of the Japanese," he wrote. "In the residential area of the Bronx where I lived, there were many elderly couples living by themselves, and I often saw ambulances going to take an elderly person who had suffered a heart attack to the hospital."

Endo was linking three things—diet, fat, and heart disease—just like many other medical experts of the day. Doctors knew that a lot of their heart patients had a fatty buildup clogging their arteries, slowing the flow of blood to the heart. When they looked at those arteries more closely, they saw that the buildup was usually made up, in large part, of cholesterol. Studies had shown links between levels of cholesterol circulating in the blood and the development of heart disease, and between diets high in saturated fats (the kind you get from fatty meats, dairy foods, and lard) and levels of blood cholesterol. A picture emerged: A diet high in saturated fats led to high blood cholesterol, which led to clogged arteries, which led to heart attacks.

If that was true, you didn't want your cholesterol too high. But you also didn't want it too low. Cholesterol in the right amounts is vital for health. You find it everywhere in the body, in every organ, and it's a central component of every cell membrane, including the lining of nerve cells. A lot of your brain is cholesterol. Your body also uses it to produce other needed bits and pieces, from vitamin D to bile acids. It's absolutely essential, so your body makes a lot of it: Three-quarters of the cholesterol your body needs is produced in your liver. The rest comes from diet.

It was the diet part that was being linked to heart disease. And heart disease was—and is—America's biggest killer. The years around 1960 were the high-tide mark for American heart problems, and death rates were through the roof. Maybe it was the smoking, maybe the drinking, maybe it was stress or sitting in front of the television or all those desk jobs. And maybe the culprit was fatty, cholesterol-rich foods.

If it *was* high cholesterol that was to blame, Endo thought, maybe molds made a medicine that could fight it. A wonder drug to lower cholesterol. Something like a penicillin for heart disease.

After returning to Tokyo and getting a job at a drug research firm, he started searching. Endo collected one fungus after another, growing molds in his lab, then testing the soup of chemicals they produced. He went through almost four thousand different species before he found what he was looking for.

It happened in 1972. The winner was a blue-green mold Endo found spoiling a bag of rice in the back room of a Kyoto grain shop. It turned out, oddly enough, to be a kind of *Penicillium*. He found that this one made a chemical that dramatically affected cholesterol levels. It looked like exactly what Endo had been searching for. And as he spent months purifying and testing it, his excitement grew. It was, as he put it later, "extremely potent."

It worked, he found, by blocking the body's ability to make its own cholesterol, shutting down an enzyme that was needed at a critical early point. Blocking this enzyme (HMG-CoA reductase) was something like throwing a monkey wrench into a machine at the start of an assembly line. Given this drug, cholesterol levels in the blood fell. Even better, it turned out that the body, trying to adjust to the decline in cholesterol, would find more ways for cells to scavenge what was left from the blood. Endo's experimental drug not only lowered the body's production, it also *increased* cholesterol uptake by cells, giving his new drug a one-two punch.

In 1978, Endo's drug was tested on a young woman with a genetic disease that resulted in cholesterol so high that pockets of it gathered under the skin around her eyes and joints. No matter what she ate, her blood cholesterol was four times that of most people. Many in her family had died of heart disease, and it was almost certain she would, too.

Endo's drug lowered her blood cholesterol levels by 30 percent within a few days. But then she started experiencing side effects, including aches, pains, and a weakness and wasting of her muscles. She was taken off the drug for a while, then the researchers tried again with lower doses. It was better this time. Tests were extended to more patients. Over the next six months, a total of eight patients with very high cholesterol levels were given the experimental drug, dropping their blood cholesterol levels significantly with no serious side effects. This was very promising. The results were published in 1980.

Everything was going so well that it came as a shock to Endo when his company told him it was shutting his program down. A more serious side effect had come up in another laboratory, where toxicity tests were being done on animals. It looked like a group of dogs dosed with the drug had developed a kind of blood cancer. And the hint of cancer in test animals was all it took. The company pulled the plug.

Endo thought it was a mistake. The dogs in question had been given what he recalled as "astonishingly high doses" of the drug, about two hundred times more, pound for pound, than any human would ever receive. There was even some doubt that the test animals developed cancer at all (and indeed, later studies showed that the dogs probably didn't have cancer, but instead suffered a buildup of treatment-related waste that was mistaken for cancer).

It didn't matter. The risks of Endo's drug were deemed too high. The Japanese halted development. Endo's pioneering effort came to an end; he would never make any money from the eventual success of the drugs he discovered.

The focus of development now moved to the United States. After it became clear that the cancer-side-effect scare was just a questionable, probably mistaken observation, drug firms dove back into the field. They found other molds that made chemicals similar to Endo's. These were tinkered with chemically to make even more variations. They all worked on the same enzyme, had roughly similar cholesterol-lowering effects, and looked to be surprisingly safe. These were the first statins.

THE TIMING WAS RIGHT, and the potential profits were staggering. Just as Endo had noticed that Americans tended to be fat and also had a lot of heart attacks, other researchers were gathering evidence that the main cause of heart attacks—the deposits that built up and clogged the vessels around the heart—also seemed to be linked to high cholesterol. What was the relationship?

A clue came from the laboratory of Russian researcher Nikolai Anitschkow in the years just before World War I. In these waning days of the empire of Tsar Nicholas II, Anitschkow, neatly groomed and precisely dressed, was trying to discover what caused the arteries of older people to thicken and harden. Most doctors considered it a natural and inevitable part of aging. Anitschkow believed it was related to diet. So he started feeding rabbits high-fat foods and injecting them with cholesterol, looking for signs of heart disease. He found that in his laboratory he could reliably produce fatty deposits very much like those found in human heart patients in the arteries of rabbits. He thought he had found the key to the hardening of arteries.

Critics jumped on his experiments, pointing out that of course he was making rabbits sick with his high-fat diets—they were herbivores, after all, so such a diet was unnatural. Humans were not herbivores. When he reran his tests on dogs, he found that he couldn't produce the same results. But when he used chickens—like humans, chickens are omnivores—he could again produce the fatty buildup in arteries.

Scientists argued about his results for decades, kept doing experiments, and views gradually shifted toward linking heart problems with fat and cholesterol.

The man who put it all together—at least in the public's mind—was Ancel Keys, a Minnesota researcher who for decades from the 1940s through the 1980s promoted the idea that heart disease and cholesterol levels were inextricably linked, and that controlling dietary cholesterol could dramatically lower the chances of a heart attack. Ironically, some of his most compelling evidence came from looking at diets in Japan, where people ate much less saturated fat and suffered far less heart disease. More support came from huge population analyses like the Framingham Heart Study in the 1950s, which identified cholesterol and high blood pressure as the two leading prepathological markers for people at risk of heart disease. Put in its simplest form, Keys's work (and that of many other researchers) could be summarized as this: High-fat diets lead to high serum cholesterol levels, which increase the risk of heart disease. (Serum cholesterol is a total measure of all types of cholesterol in blood, including "bad cholesterol" or LDL, "good cholesterol" or HDL, and triglycerides.)

We now know that that description is too simple (although most of the public and much of the health-care community still take it as gospel). The links between dietary fat, serum cholesterol, and heart disease are more complex and subtle than early researchers thought. If you mapped all the links, they'd look less like a straight line and more like a bowl of spaghetti—lots of strings, loops, and tangles. Then there are some simple, confounding facts: People with low cholesterol sometimes get heart disease; plenty of people with high cholesterol never develop heart disease. High cholesterol, it turns out, does not cause heart disease like a germ causes an epidemic. It is, instead, a risk factor—one among many.

And that is an important distinction. We are accustomed to thinking about disease as springing from a single cause, like one type of bacteria causing one kind of infection, or one type of chemical causing cancer, or a deficiency in one vitamin causing a problem. We've still got a one-culprit-per-disease mind-set, with the linked thought that once we find the culprit, we'll find the medicine to stop it. In the last half of the twentieth century, cholesterol became more or less that culprit for hardening of the arteries and heart disease. Once we identified the culprit, all we needed was a magic bullet to kill it.

Yes, many diseases—especially infectious diseases caused by viruses, bacteria, and parasites—have a single cause, a well-defined target we can aim for. Those are the relatively easy targets, the ones that we began to hit with the smallpox vaccine (page 65) and sulfa drugs (page 108). As these one-target contagious diseases began to fall, one by one, to antibiotics and vaccines, medical researchers got into more difficult, more tangled territory. Now the big killers in the United States are cancers, heart disease, stroke, lung problems like emphysema (usually linked to smoking), diabetes, and, increasingly, Alzheimer's. Apart, perhaps, from the simple advice to stop smoking, there is no easy answer, no miracle drug, no magic bullet for any of them. They all have multiple, often little-understood causes. They arise from a complicated web of factors, some genetic, some environmental, some general, some personal, that add up to disease in ways that we are still struggling to understand. Because of the complexity of these diseases and the number of unknowns involved, we talk about risk factors—habits and exposures that might shift the chance of getting a disease one way or another—more than root causes. This is the new reality of medicine today, as we start our assault on the last great killers, the toughest health challenges we have yet faced.

But back in the 1980s, it seemed like cholesterol was the sort of clear, well-defined enemy that we were used to fighting. Tackling cholesterol would help us unclog arteries and reduce the number of deaths from heart disease. It was a simple approach to a complex problem.

Perhaps too simple. The National Academy of Sciences released a report in 1980 suggesting that widespread efforts to control cholesterol levels lacked a good scientific basis, and many researchers remained unconvinced that cholesterol was all that bad. Regardless, the public, spurred by their physicians, started getting their cholesterol checked and making lifestyle decisions based on the results. By the mid-1980s cholesterol levels were carefully tracked, lowering cholesterol became a national priority, and the age of low-fat diet fads was upon us.

It was the perfect moment for statins. Drug companies poured millions into developing and testing variations on Endo's theme, and they started hitting the market. Merck was the first to the finish line, gaining approval for lovastatin (trade name Mevacor), in 1987. It was quickly joined by similar products from other companies: simvastatin (Zocor), pravastatin (Pravachol), atorvastatin (Lipitor), fluvastatin (Lescol), and current bestseller rosuvastatin (Crestor). Within a few years, it seemed that every big drug company was selling one.

Physicians loved them. Statins quickly became blockbuster bestsellers, combining safety and reliable decreases in serum cholesterol with that all-important factor, timing. They hit the market just as middle-aged baby boomers were beginning to look askance at their fast-food diets and expanding waistlines, and public concerns over high cholesterol were peaking. At first, they were prescribed to patients with very high cholesterol levels and a strong family history of heart problems. But once they were approved for sale, drugmakers threw millions of dollars into additional testing to show why their particular brand was better than their competitors' and to expand the market by seeing if the drugs might also be useful for lower-risk patients. They found small but real benefits for preventing heart problems in more and more people with lower and lower risks. Every new study showing a positive effect was highly publicized.

The whole thing snowballed. Worries about cholesterol fed the market for statins, and the research on statins fed the worries about cholesterol. And the whole thing was fueled by the diet industry, with an excruciating amount of attention paid to what people were eating. Suddenly a desire for French fries and ice cream was no longer a personal choice. It was a recipe for disease, with pressure from both drugmakers and diet faddists moving millions more people toward worrying about the cholesterol in their blood. As one expert put it, "Interest in a medical condition tends to increase in tandem with the development of its drug. . . . The drug transforms a bodily state into a treatment category and then into a disease category."

Just as high cholesterol became established as a health risk in the public mind (with the definition of "high cholesterol" constantly moving downward thanks to a steady stream of research reports funded by statin makers), statins appeared on the scene to treat it. The result was incredible sales. One statin alone, Lipitor, became the most commercially successful drug in history, with sales between 1996 and 2011 that topped $120 billion. All statins combined are expected to generate more than $1 trillion per year in sales by 2020—more than the annual GDPs of all but a handful of nations.

As drug companies funded study after study showing marginal benefits for more and more patients, heart specialists and heart disease foundations jumped on board. The old skepticism about the role of cholesterol and its control in heart disease—like an Office of Technology Assessment report released during the early statin years that estimated widespread use of the drug could cost society somewhere between $3 and $14 billion a year, with unclear benefits and a cost of $150,000 per life-year saved—melted away before an onslaught of drugmaker-funded studies, drugmaker-supported conferences, and the enthusiasm of medical experts, many of whom had financial ties to drugmakers. The many ways in which drugmakers influence researchers, health-care providers, foundations, government agencies, and the public—the ways in which they shape modern health care—is a fascinating story. And at its heart, it's not terrifically complicated.

Put simply, today's large drug companies are great at finding evidence for therapies that promise profits, pretty good at downplaying evidence that gets in the way, and grand masters at promoting their products to physicians and the public. Some critics paint drug companies as evil masterminds—"Big Pharma"—out to ruin our health in order to line their pockets. I don't see it that way. But I do recognize great businesses when I see them, and today's big drug companies are often brilliant at what they do, from cutting-edge research and development to highly effective marketing and advertising. I recognize that drug companies are private corporations; their primary responsibility is to generate profits for shareholders. And they are, in general, very good at that. Yes, they do push the boundaries sometimes, especially when it comes to making people feel like they should take a new drug to treat what might be a marginal condition, when it comes to extending patent protection, raising prices on some drugs, or when it comes to persuading physicians to prescribe their products. We need good watchdogging from public agencies like the FDA and continued attention to making strong drug laws. With adequate public scrutiny in place, I have no great worries about Big Pharma (although I do wish that the public knew more about the business, so they could make better-informed decisions about which drugs to take). Readers who want to know more about this intricate big-money dance should read medical historian Jeremy A. Greene's cool and convincing book *Prescribing by Numbers*.

For statins it came down to this: A growing consensus in the 1990s and early 2000s, fueled by generally well-run, often industry-funded research, showing that statins were useful in staving off heart disease for more and more patients at lower and lower levels of risk. The benefits might be very small, but they were there. A few enthusiasts—only half joking—recommended putting statins into the water supply.

So that, I thought, is why I got my form letter. I'm in my early sixties (a risk factor in itself) and I've got somewhat elevated cholesterol. My heart had always been fine, blood pressure normal; I don't smoke, I get moderate exercise, I have a pretty good diet, and I've never experienced a heart problem. Twenty years ago, I did have what is called, hilariously, a "cerebrovascular accident"—a tiny blood clot temporarily blocked blood flow to a part of my brain responsible for my sense of balance. After a few hours of vertigo and some blood thinners in the hospital, it went away without any long-term effects. That went on my record as a heart-related risk factor. And today that little blood clot, plus my elevated cholesterol reading, led some computer program to tell some faceless experts in my local health-care system that my risk factors were high enough to warrant taking a statin. It was all numbers being crunched and form letters being pumped out. It was health care by algorithm. The result: a physician I've never met recommending that I consider taking a new prescription drug, potentially for the rest of my life.

This is a recent and remarkable change in the practice of medicine. We are, as a society, moving beyond the idea of health built around how we feel as individuals and toward a world in which our care is determined by our position on a statistical curve. In my case I *feel* fine, but my numbers aren't right. When the numbers aren't right, you run a higher risk of some future heart-related problem. Taking a medicine to lower cholesterol, the reasoning goes, will lower the risk.

Doesn't sound too bad when you put it that way.

So why did that form letter tick me off? Because I don't want my health decisions divorced from how I feel. I don't want computers determining my health-care recommendations instead of my personal doctor. I'm one of those throwbacks who wants to be treated like an individual human, not a set of data points.

Before making a decision about taking a statin, I needed to know more about my personal chances of benefiting from the drug, and how much risk I would actually run. So I did what science guys like me always do: I sat down at my computer. I had questions, and I thought the Internet could answer them: A statin was going to give me some benefit, but how much? There were small risks, but how many? How much should I worry about my personal risk of heart disease? I started putting together a simple risk/benefit analysis, pros on one side, cons on the other.

Benefits versus side effects. Sounds easy enough. But the deeper I dove into understanding statins, the more complicated things got.

LOWERING CHOLESTEROL is the benefit, right?

Well, not exactly. The real benefit, the thing everyone is shooting for, is avoiding heart problems. *That's* the goal. Many physicians (and every drug company making a statin) believe that statins do the trick. And in many cases—especially those with patients who have really high cholesterol and a history of heart problems—they do. For high-risk heart patients, statins are undisputed lifesavers.

But it gets less clear-cut with people like me, moderate-risk patients with elevated cholesterol (but not red-lights-flashing high) and little or no family or personal history of heart disease.

My research led me quickly to the old Ancel Keys lipid hypothesis, and the whole idea that dietary fat leads to high blood cholesterol and heart disease. I took that hypothesis for granted; I grew up with it. I thought it had been proven true in the 1980s and 1990s.

But the more I read about the lipid hypothesis, the more dubious it looked. For one thing, all those low-fat diets didn't end up doing as much good as people thought they would. As expected, a lot of people found that eating less fat in their diet could lower their serum cholesterol levels. But along with their low-fat diets, many Americans switched their foods to those richer in sugars and grains, which pushed up diabetes rates. Diabetes is a risk factor for heart disease. And, in general, the more added sugar people ate, the higher their chances of cardiovascular disease. So it was hard to untangle the effects of a low-fat diet by looking at the resulting rates of heart disease in the real world.

There was another confusing thing, too: Heart disease rates had peaked in the 1950s in the United States and had begun falling in the early 1960s, decades before statins were available. A lot of that had to do with lower rates of smoking (another major risk factor for heart disease). And the rates kept going down after statins. But changing the nation's attitude toward fats and adding all those drugs didn't change the trajectory much.

Many researchers studying the relationship between cholesterol, statins, and heart disease were also puzzled. As studies continued, they found themselves coming to grips with a welter of confusing, unexpected, and paradoxical findings. Statins are among the most-studied drugs in history; you would think that after decades of intensive research and the use of truckloads of cholesterol-lowering drugs by millions of patients, we'd be able to settle the whole question of how diet and drugs relate to blood cholesterol levels, and how that all affects heart disease. But that relationship remains murky, with a large and growing literature questioning any simple answers.

For instance: In one 2016 study, researchers followed more than 31,000 patients taking statins, tracking their LDL cholesterol levels (LDL is the infamous "bad cholesterol") and incidence of heart disease. They found that lowering very high LDL levels did help prevent heart disease—but only to a point. Surprisingly, they found that patients who got their LDL down to the lowest levels—below 70 mg/dL, the target for many statin regimens—did no better than patients who only got them down between 70 and 100. In fact, anything below about 90 seemed to do nothing additional to prevent heart attacks. Lower cholesterol was not necessarily better. A strike against the lipid hypothesis.

In another 2016 paper, an analysis of nineteen studies concluded that the evidence showed that lower LDL cholesterol didn't seem to do much to lower overall mortality (that is, death from all causes) in patients over sixty years of age. Worse than that, as LDL levels went down, cardiovascular mortality actually went *up*. There were even hints that higher total cholesterol in the blood might somehow be protective against cancer. "Since elderly people with high LDL-C [total LDL cholesterol] live as long or longer than those with low LDL-C, our analysis provides reason to question the validity of the cholesterol hypothesis," the authors concluded.

And yet another recent systematic review of forty studies concluded, "Dietary cholesterol was not statistically significantly associated with any coronary artery disease," even though it might raise total blood cholesterol. What about statins? As expected, many studies pointed out their benefits. But others found their benefits marginal or nonexistent. As a 2015 scientific review of major statin studies summarized: "A careful examination of the most recent statin randomized clinical trials . . . clearly shows that contrary to what has been claimed for decades, statins do not have a significant effect in primary and secondary prevention of cardiovascular disease."

There are an equal number of studies that argue that statins do, indeed, lower the risks of heart disease for many moderate-risk patients, so the scientific back-and-forth continues. And that's expected: Science at its best is a series of arguments over the validity of data. Scientists are chronic skeptics of one anothers' work, and they should be, because it's only out of careful criticism, constant argument, and repeated study that strong facts emerge.

Given the state of statin research, my take-home was this: In general, very high blood cholesterol correlates with a higher risk of heart problems. It is a risk factor. But it's a complex risk factor with a lot of caveats and sometimes debatable effects. And it's only one among many, with smoking, family history, diet, and exercise playing just as large a role. Statins are great for patients with very high blood cholesterol, especially when it's linked to a family history of high cholesterol—the groups they were first approved to treat. But for people like me, moderate- and low-risk patients with somewhat elevated cholesterol, the benefit of taking a statin remains, at best, debatable.

YOU WOULDN'T KNOW THAT, however, from reading statin ads. A few years ago, for instance, a magazine ad for Lipitor (a bestselling statin) led with this bold headline: "Lipitor reduces risk of heart attack by 36%*."

That certainly sounds good. But it also seems out of whack with what I'd been reading about the benefits of statins. So I followed the asterisk. It led to some much smaller type at the bottom of the ad: "*That means in a large clinical study, 3% of patients taking a sugar pill or placebo had a heart attack compared to 2% of patients taking Lipitor."

Do a little deciphering, and you'll find that this is what that ad really says:

Take two hundred people with risk factors for heart disease and randomly break them into two groups of one hundred patients each. One group gets a daily statin, the other gets a placebo (a do-nothing pill that looks like a drug but isn't one). Now track what happens. After some time—six months, a few years, whatever the length of the study—you count how many people in each group have had a heart problem. You find out that in the placebo group there were three heart attacks. And the statin group had only two heart attacks. The statin works! It appears to have prevented one heart attack.

But how do you communicate that to the public? You can't do it as I just did in the previous paragraph, because that explanation takes too long and seems too weak. You have to boil it down to something simpler and stronger. So you look at the numbers a certain way. Drug companies like to emphasize what's called "relative risk," because it tends to make benefits look bigger. In this example, the placebo group had three heart attacks, the statin group had two. If you're just looking at those few patients who had heart attacks, you've lowered the risk about a third, from three to two. A 33 percent reduction in heart attacks! Cue the ad writers.

The number is both true and misleading. Relative risk looks only at the small number of patients who suffered a heart attack. It ignores everyone else in the test. Remember that the vast majority of people in both test groups, whether they took the drug or not, didn't have a heart attack. For them, taking the statin didn't make any difference at all. If you look at the whole test group, not just those who got heart attacks, taking the statin prevented one heart attack in every hundred patients. This is a reduction in "absolute risk," and in this case it's 1 percent. But a headline saying, "Cuts heart attacks by 1%" doesn't sound so great. And yet it's also true. The well-paid men and women who write drug ads make their money by doing things like playing up relative risk and ignoring absolute risk.

Which is right, relative or absolute? They both are. It's just a matter of what you want to emphasize. Physicians tend to take both into account. And when you look at it that way, even a reduction of just 1 percent in absolute risk can mean preventing thousands of potentially devastating medical problems when spread over a large population. However, it also means that millions of patients might be taking a drug from which they get no benefit at all.

WITH MY FAITH in the lipid hypothesis shaken, I wanted to know more about my real risk of heart disease. And that sent me down another rabbit hole.

It turns out that estimating your personal risk of heart disease is anything but a precise science. Given that cholesterol numbers alone are apparently less predictive than they once seemed, physicians are gingerly moving away from relying on them in favor of weighing a number of risk factors.

Here are the major risk factors for heart disease:

- high blood pressure
- history of smoking
- diabetes
- high cholesterol
- age
- family and/or personal history of heart disease

By getting a patient's history and weighing risk factors like these, physicians can plug it all into a formula and guesstimate a patient's future risk of heart problems.

You can do it yourself online—there are a number of sites in which you can input your numbers and find out your odds of having a heart problem in the future—but take any results you get with a large grain of salt. If you go to enough different online heart disease risk estimators, you'll find they use somewhat different mixes of risk factors; your results may vary.

More important is what your doctor recommends based on that rough idea of risk. And things have been changing here, too. Today, physicians are more likely to prescribe you a statin than they were a decade ago, because they consider more and more of their patients to be candidates for statin therapy. Here's why:

In 2013, two highly respected organizations—the American College of Cardiology (ACC) and the American Heart Association (AHA)—put out a newsworthy new set of guidelines for prescribing statins. The new recommendations dramatically lowered the thresholds for recommending statin treatment, dropping it from a 20 percent risk of future heart disease to something closer to 7.5 percent. It expanded the pool of potential statin patients enormously. Suddenly, millions of people who had never had heart disease and were considered moderate risk were being recommended drugs. Again, my letter.

CONTROVERSY HAS FLARED ever since, with researchers arguing for and against the 2013 guidelines in the pages of medical journals, blog posts, and media stories, hashing over everything from the accuracy of risk estimates to which statin studies are most valuable. Some doctors think the ACC/AHA guidelines are gold, others think they're worse than useless. There is still no overwhelming consensus in the scientific community.

There's no doubt that if you've had a heart attack already, you're automatically a high-risk patient, and statins are pretty good at lowering the risk of a second attack. This is called secondary prevention. No question about using statins here.

But that wasn't me. I've never had a heart attack. I'm what's called a "primary prevention target"—the idea is to try and prevent a problem before it happens. And primary prevention is where the action is for statins. The new guidelines, by emphasizing the use of statins in the larger group of moderate-risk primary prevention patients, created good news for drug company shareholders—and a mixed blessing for patients. Because the more any drug is prescribed, the more people suffer from side effects. And statins, while very safe compared to most drugs, do have side effects.

NO DRUG COMES without side effects. That is true of drugs we might take every day like caffeine, drugs in our medicine cabinets like aspirin, or each and every one of the thousands of drugs available by prescription. When it comes to medicine, the rule is that you don't get good effects without some (hopefully much lower) risk of bad ones.

The most commonly reported statin side effects include:

- muscle pain and weakness
- diabetes
- memory loss, cognitive problems

Rare but more serious side effects include:

- rhabdomyolysis (severe muscle deterioration that can lead to kidney damage)
- liver injury
- Parkinson's disease
- dementia
- cancer

The risk of side effects generally goes up as the drug dose goes up, so patients taking higher amounts of statins tend to have more problems. Most physicians try to keep doses as low as possible, as long as they can achieve the desired result.

There is a lot of controversy over the most common statin side effects—both about how often they happen and how severe they are.

Muscle pain and weakness

Somewhere between one-tenth and one-third of all statin patients report some degree of muscle-related problems after they start taking the drugs. Why the big range? Partly because many large research studies ignore them, considering them too minor and too subjective to track. Doctors know that it's hard to distinguish between everyday aches and pains—the kinds that would bother someone whether they took a drug or not—from those that might be due to the drug. Some studies indicate that reports of muscle problems may be overblown, in great part the result of patients paying more attention to their bodies after starting the drugs and blaming what would otherwise be normal creaks and cramps to the medicine. There are well-documented cases where even patients taking a placebo will start experiencing side effects—this is the so-called nocebo effect—because they think they're taking a drug that might cause side effects. This makes low-level effects like muscle pain especially difficult to track. For the most part, however, statin-related muscle problems are considered minor and are generally solved by taking a temporary break from the drug, or by switching to a different statin.

At the same time, there is little doubt that real muscle pain and weakness does affect many people taking statins. It can get bad enough to affect mobility and exercise tolerance. In fact, it is the number one reason people stop taking statins. Most of the time the drug-related effects are very mild, ranging from stiffness and soreness to cramps and weakness. In very rare cases, the drugs can cause more serious problems, from crippling inflammation to life-threatening muscle damage. Some researchers even think that statins might spur heart problems by damaging muscle action in heart and blood vessel muscles, although the evidence here is weak.

Other statin researchers worry that muscle problems are signs of something bigger. After all, why should taking an anti-cholesterol drug cause muscle-related side effects at all? The answer might lie in the energy-producing centers of cells, microscopic structures called mitochondria. The idea is that statins might somehow affect mitochondria, thus leading to the reported weakness and pain. Mitochondria play a vital role in many cellular functions; in fact, we can't live without them. The possibility of statin-caused damage to mitochondria—which could have long-term effects that go far beyond aches and pains—is now being investigated at a number of research centers.

Diabetes

Most physicians don't worry much about minor aches and pains. But they do worry about the link between statins and diabetes. Here again, there is argument and controversy over just how serious a problem it is. Most early statin enthusiasts dismissed the danger entirely. But more recent, longer-term studies are showing that a small increased risk of diabetes is real.

While it's now widely accepted that statins boost the risk of diabetes, questions remain about how much. At one end of the spectrum are studies showing that taking a statin for a year or more increases the occurrence of diabetes somewhere between a low of four or five new cases for every thousand people taking the drug to a high of five or six times higher. One large review concluded that taking statins would spur diabetes in about one patient in one hundred. It depends on the study, the drug dose, and the amount of time the patients are tracked, as well as the diabetes risk that the patient ran before the statin was started. The higher your pre-statin risk of diabetes, the more starting a statin tends to boost its onset—as if statins were unmasking diabetes in those most at risk of developing it anyway. As a paper written by physicians at Johns Hopkins recommended, "People with pre-diabetes should only be treated with statins if they have a markedly elevated risk of heart attack and stroke."

The jury on diabetes is still out in part because most studies are relatively short-term, running no more than a few years. Longer studies are needed to gauge the full risks of potentially long-gestating conditions like statin-related diabetes. We're likely to hear more about this in coming years.

Cognitive problems

No side effect of statins is less well understood than patient reports of memory loss, confusion, "brain fog," and assorted other problems with brain function. The effects are mostly mild and tend to go away when statins are stopped. Like minor muscle pain, they're hard to track or definitively link to statin use. Most early studies did not even monitor these hard-to-pin-down side effects, and most doctors don't consider them important enough to worry about. But anecdotal reports have been common enough to spur the FDA to add a warning about cognitive side effects in all statin-labeling information.

One thing just about everyone agrees on is that we need more information about statin side effects. It's important to remember that these drugs, overall, are among the safest ever developed. Compare statin side effects to those carried by drugs that few people worry about—aspirin, for instance, with its risks of ulcers, cramping, and internal bleeding, is a drug that kills thousands of people every year—and you begin to understand just how minor the downsides of statins are.

But there is reason to believe that most studies to date have tended to downplay the extent of statin side effects. In part that's because most of the side effects are so mild that they fall below the threshold of concern for physicians. And in part it's because most studies have been done by or supported by drugmakers, and in their communications drugmakers tend to highlight benefits and minimize risks. Another factor to keep in mind: Many side effects can take years to surface, and most results to date have come from short-term studies.

If statins follow the pattern of many other big-selling drugs, we will find out more about the true size of benefits and side effects as time goes on, as more people take statins, and as longer-term studies are completed. One thing we can be sure of, as *Scientific American* put it, is that "There is no doubt that the rising number of statin users will be associated with increased reports of negative side effects."

It's that Seige cycle again. We're beyond the stage 1 honeymoon for statins and into the stage 2 period of more critical review. After longer-term independent work provides a fuller, more balanced picture, we'll reach stage 3, and statins, like all one-time miracle drugs, will be seen for what they are: important additions to care in some cases, unnecessary in others.

THE GROWING USE of statins brings up two other related, larger, and more-or-less-hidden issues.

One is around the "medicalization" of our lives. This somewhat ill-defined term is used to describe a troubling trend in our society, in which things that we once simply dealt with on our own—like lifestyle choices, low-risk health conditions, personality quirks—are now being turned into treatable medical conditions. Often this goes hand in hand with the appearance of a new drug suited to treating the new condition. Tranquilizers are a classic example. When Miltown, the first minor tranquilizer, was discovered around 1950 (see page 148), nobody quite knew what to do with it. There had never been a drug for minor anxiety. It was considered an unimportant problem; sufferers handled it on their own, talked it out with friends or advisers, and waited for it to pass. But when a drug appeared to treat it, suddenly minor anxiety became a drug-treatable condition. It was rethought, redefined, and medicalized, and tranquilizers became blockbuster drugs. Somewhat the same thing happened when drugs for ADHD became available—what had once been considered behavioral problems in school became a drug-treatable disease, with definitions of who might benefit broadened and broadened until it seemed every tenth kid was taking some kind of drug. This expansion of treatable disease categories might be well-intentioned, but it's also a bit scary. The universe of conditions that might get some benefit from a prescription drug balloons until millions more people think they or their loved ones are sick or dangerously at risk—diagnosably, drug-treatably at risk—even if they feel fine. Minor problems can become major moneymakers for drug companies. With a much larger group of potential patients getting increasingly worried about their risks, the market for the drug grows. Blockbusters can result.

At best, medicalization is an attempt to improve health by recognizing how the power of modern healthcare can be applied to a wider and wider range of problems and used to head them off before they get worse. At worst, it can turn into what's called "disease mongering"—emphasizing or redefining the risks of disease to enlarge the market for medications.

Are statins part of this problem? Some statin critics charge that broadening the patient base to include tens of millions of seemingly healthy people—most of them middle-aged, with some degree of heightened risk but without any history of heart problems—is another way to medicalize our lives, to get people with no symptoms of disease to start taking a drug. There are strong counterarguments here, too, with proponents of wider statin use pointing to the need for the drugs to counter the effects of ever-richer diets and ever-more-sedentary lifestyles.

This is an area of active debate. But the result at present is that more and more people are being prescribed statins to prevent a shrinking number of heart attacks.

This leads to a second, somewhat hidden side effect: the use of statins as a way to avoid tougher personal choices. By taking a statin, users might think that their cholesterol problem is solved and sidestep more difficult lifestyle changes related to diet and exercise. Some researchers are concerned that the drugs offer false reassurance: the idea that taking a statin can make up for poor dietary choices and a sedentary life. Taking the pill solves the problem, and means you don't have to exercise as hard or eat as many vegetables. Or, as one medical expert put it, drugs like statins "short-circuited the link between effort, responsibility, and reward in the arena of health."

And there's some evidence that it's happening. For instance, a 2014 study (with the decidedly unscientific subtitle "Gluttony in the Time of Statins?") found that patients on statins were significantly more likely to increase their fat intake and calories than nonusers—and put on more weight as a result. The trend has worsened over the past decade. "We need to consider if it is an acceptable public health strategy to encourage statin use without also taking measures to decrease the likelihood that its use will be associated with increased caloric and fat intake as well as weight gain," the authors concluded. "We believe that the goal of statin treatment, as with any pharmacotherapy, should be to allow patients to decrease risks that cannot be decreased without medication, not to empower them to put butter on their steaks."

The key, experts agree, is to emphasize heart-healthy diets and moderate exercise—even when you're taking a statin.

AS A RESULT of getting a piece of junk mail I didn't like, I spent months wading through stacks of articles, books, and editorials about statins, making myself into a better-informed patient. I came away with a better understanding of the drugs.

And now I've finished my personal risk-benefit analysis. For patients like me—low to moderate risk, never had a heart problem, but have some markers for increased risk—according to the best data I could gather, the data looks like this:

- Between 100 and 200 people at my risk level would have to take a statin for five years to prevent one fatal heart attack

- Between 150 and 270 people would have to take a statin for five years to prevent one stroke
- Around 50 to 100 people would have to take a statin for five years to prevent any sort of cardiovascular problem (both fatal and nonfatal)

And risks? Ignoring all the rare side effects:

- If I started a statin, I would have a one-in-ten chance of developing some degree of minor muscle problem
- I would run an increased risk of developing diabetes— about the same as my decreased chance of having a fatal heart attack

Things were getting clearer. But it still wasn't *crystal* clear. I ended up pretty much where the authors of a recent overview of statins for low- and moderate-risk patients did when they wrote: "It is likely that the benefits of statins are greater than potential short-term harms although long-term effects (over decades) remain unknown. Caution should be taken in prescribing statins for primary prevention among people at low cardiovascular risk."

WELL, THERE COMES a point where you have to make a decision. So I made mine. After talking it over with my primary care physician—a gregarious guy who recommended that I "blow the rust out of my pipes" with a statin (I love the way my doctor communicates medical information)—I told him no, there was not going to be any rust-blowing. What I will do instead is to keep a better eye on my diet and exercise. Nothing radical. I'll also write a nice letter to my healthcare provider asking them to stop sending me unwanted advice. I will watch drug ads with a new skepticism. I will back-burner any worries about heart problems. I will forget about statins and concentrate on enjoying life.

But that's me. Others in roughly the same risk category, reading the same information about the same drugs, might respond differently. Some will simply do what their doctor recommends. Some will see it like the lottery: Your chances of winning might be low, but you can't win if you don't buy a ticket. So they'll take the statin to prevent that one-in-a-hundred heart attack. Risk-averse types might look at it like an insurance policy: The chances of a bad event are small, but it's better to be covered just in case. Millions of people take statins with no problems.

And that's fine, too. If you can afford it, if you accept the potential side effects, if you keep up your exercise, and if you control the urge to slather your steaks in butter, go for it.

But it's not for me.

CHAPTER 10

A PERFECTION OF BLOOD

IF STATINS (PAGE 211)—a case study of the power marketing can hold over medicine—exemplify some of the worst of today's Big Pharma, this next story offers an antidote. The discoveries here flowed out of old-style dedication, scientific altruism, and generous friendship. It all resulted in a gift to the world: a large and growing family of drugs that are so precise, so powerful, and so safe, they're changing the ways we think about medicine.

The term *monoclonal antibodies* sounds intimidating until you break it down. *Mono* means one, like in *monogamy*. *Clonal* is about making clones—exact genetic copies of an original, like Barbra Streisand's dogs. And antibodies are the infection-fighting molecules that white blood cells give off when they're fighting off invaders. Antibodies are like guided missiles in the blood, able to recognize and lock onto germs and viruses, then helping to clear them from your system. So there you have it: Monoclonal antibodies are guided missiles produced by clones of identical white blood cells.

Why are they a big deal? Because monoclonal antibodies are the closest thing we have to true magic bullets. Half of today's top-ten bestselling drugs are monoclonal antibodies. You can recognize them by their scientific names, which all end in *mab* (for Monoclonal AntiBody). They include infliximab (trade name Remicade), for autoimmune diseases; bevacizumab (Avastin) for cancer treatment; trastuzumab (Herceptin) for breast cancer; rituximab (Rituxan) for cancer. At the very top is adalimumab (Humira), which is being used for treating a growing list of diseases linked to inflammation. These drugs are making billions of dollars.

And many more monoclonals are coming.

All of these things—clones of white blood cells, antibodies directed toward specific diseases, big money—are linked to a bewilderingly complex and absolutely vital part of your body: the immune system. When I was in school back in the late 1970s, we knew a lot less about it, and to me the immune system looked like something Rube Goldberg might have made if he was high on acid. There were far too many players, it seemed to me—a baroquely elaborate, stupefyingly tangled web of organs and cells and receptors and antibodies and signals and pathways and feedback and genes and enzyme cascades that somehow all work together to keep you safe. Today we know much more, and the immune system now looks more like a symphony orchestra, with each player making different noises but all playing the same composition, creating a grand piece of music.

The immune system somehow knows how to distinguish what is you—your own cells—from what isn't. It has the ability not only to recognize billions of different foreign substances, but it can then direct white blood cells to make millions of antibodies, each one meticulously designed to latch onto a specific target. The system then remembers each of these not-you invaders for years, even decades. This is how Lady Mary Montagu's inoculations worked: Exposure of a patient to a small amount of an invading substance primes the immune system to recognize and remember the invader. Years later, when another infection happens, the body can ramp up an immune response much faster than it could without that initial exposure. Result: You're protected.

But how can the cells have memories? How can they recognize invaders and distinguish between you and not you? How can the immune system respond to just about everything in nature that's not you—including millions of synthetic chemicals that have never existed in nature? We're peeling back the layers of this remarkable system, learning the immune system's deeper secrets, but much of it remains profoundly puzzling and endlessly fascinating. No wonder it has captured the attention of generations of scientists.

The real surprise is that, most of the time, it works astonishingly well. It's by no means faultless—there are autoimmune diseases, when the immune system decides that your own cells are invaders and mounts a defense; there are allergies when it overresponds to something as nonself; there are tricks that viruses and cancer cells have figured out to fool the system—but it comes close. It's in full surveillance mode right now, working intently inside of you, silently scouting for invaders, setting up defenses, clearing your system, and keeping you healthy. Most of the important parts of the system had been identified by the middle of the twentieth century, and scientists were beginning to see how they worked together at the molecular level, learning how diseases turned the system on and how things could go wrong. But one thing eluded them. All that new understanding didn't produce much in the way of effective medicines.

Until 1975.

CÉSAR MILSTEIN was the epitome of a global scientist. Born in Argentina, schooled in Great Britain, devoted to building science in developing nations around the world, Milstein seemed like living proof that science was built upon open communication and international cooperation. "Science knows no borders"—that sort of thing. This seems charmingly old-fashioned now. But Milstein was a charmingly old-fashioned scientist.

He looked the part: slight, balding, somewhat owlish behind big glasses, wearing a dress shirt and slacks topped with a lab coat. But in one important way he broke the science-nerd stereotype. He loved people. He smiled a lot. He talked a lot. He was "a man held in great affection by many," as one of his many admirers remembered, "with a special gift for friendship."

He was also brilliant in the lab—he worked at the University of Cambridge—where he focused on antibodies, those guided-missile proteins made by white blood cells. Milstein, like many other researchers, was puzzled by the sheer variety and incredible sensitivity of antibodies. The body seemed capable of tailoring an almost infinite number of different antibodies, each one precisely made to fit a particular part of an invading substance. Those targets could range from a few atoms on the coat of a virus to never-before-seen synthetic molecules fresh from the lab. The targeting was incredibly precise; exposure to a single kind of bacteria could spur an animal's immune system to raise a couple of hundred different kinds of antibodies, each one targeted to different sets of just a few atoms on the invader's surface. How was all this variety possible?

Milstein was deep into that question and many others, working with the immune system down at the level of individual molecules, trying to figure out how white blood cells could make so many different antibodies to so many different substances. You have billions of antibody-producing white blood cells in your body (these are called B cells), and once they're switched on, each one can pump out millions of antibody molecules per minute. Each individual B cell produces only one specifically targeted form of antibody. But your body has billions of B cells, so you can make antibodies to billions of targets.

Antibodies are proteins, big and complicated molecules, far bigger than most drugs (the older drugs, the ones that most chemists made before 1975, are now called "small-molecule drugs"). Antibody molecules are shaped like the letter Y, and the ends of the two arms at the top are where the antibody latches onto an invader. These sticky ends are precisely made to fit some bit of the invader like a tight handshake. The fit has to be very exact for them to stick. A few atoms' difference can ruin the link. Once the connection is made, however, it triggers other parts of the immune system and, bingo, the invader is cleared.

Milstein's lab was working to understand how the body makes antibodies with that degree of precision, and his team was looking for ways to grow B cells outside the body so they could be studied more closely. That led them to antibody-producing cells that were cancerous—myeloma cells—because while normal white blood cells stop reproducing and die out after a short time outside the body, cancerous cells can keep growing forever. They don't know when to stop—that's what makes them cancer. It also makes them great for lab studies, because if you're careful you can grow them forever in bottles full of nutrients.

At a scientific meeting in 1973, the gregarious Milstein was approached by a young German scientist who'd just gotten his PhD and was interested in working in Milstein's lab. His name was Georges Köhler. The older scientist and the younger postdoc hit it off. Their conversation turned into an invitation for Köhler to come to Milstein's lab at Cambridge, and that invitation turned into a friendship.

They seemed unlikely buddies. It was more than a difference in age—Köhler was twenty years younger than Milstein—there was also a difference in style. Milstein was straight out of the 1950s, short-haired, neatly dressed, small—he came up to Köhler's shoulder—while the German was a laid-back seventies-style hippie with a full beard and jeans. Milstein worked long hours, and postdocs like Köhler are supposed to do the same, working weekends, nights, whatever it takes to impress their lab bosses and start making their reputations. Köhler, by contrast, a colleague noted, tended toward the "languid," frequently taking time off from the lab to kick back, teach himself how to play the piano, and go on four-week vacations with his kids in a VW van.

Milstein rolled with that. He believed that true creativity, in science and elsewhere, required time for reflection. Some of the best ideas come when you're on vacation. Besides, he and the young German were now more than colleagues, spending time with each others' families, visiting each others' houses. They were an odd couple, yes, but compatible, passionate about their shared research, fond of bouncing ideas off each other. They were friends.

Köhler played around with Milstein's cancerous, antibody-producing myeloma cells, trying to get them to do different tricks that might cast light on the workings of the immune system. He learned how to fuse two different myeloma cells together, joining their DNA, as a way to explore the links between genes and antibodies. Myeloma cells were great in some ways: They grew forever and produced lots of antibodies. But they were terribly flawed in others. One great drawback was that you never knew exactly what antibodies they were producing, what precisely they were targeted to. It could be any of a billion things. These cancer cells were pulled out of mice or rats because they produced antibodies, but they could be antibodies to anything. A lot more could be done with them if researchers could match the antibodies from myeloma cells with their specific targets. Köhler tried to figure out a way to do that, but failed.

Then, sometime around Christmas 1974, he and Milstein came up with a bright idea. Instead of fusing two myeloma cells together, how about trying to fuse a long-lived myeloma cell with a normal, noncancerous mouse white blood cell? If you could get such a hybrid to live forever, like a myeloma, and it produced the specific antibody from the normal mouse cell (and you could up your odds of this by priming the mouse to make a lot of white blood cells to a specific target ahead of time), you'd have just what they were looking for: flasks of cancerous cells that all produced precisely tailored antibodies to the same known target.

No one had ever tried it before, probably because no one thought it could be done. The fusion between a cancer cell and a normal cell would likely not work, or if it did, the chromosomes from one might not work well with the chromosomes from the other, the resulting cells would be a genetic mess and probably die, or if they lived they might not make the targeted antibodies. But nothing ventured, nothing gained. Köhler gave it a shot.

He got some cells to fuse, and, as expected, most of the resulting hybrids died. But a few did not. They began to grow and to multiply. And Köhler worked with these tiny clumps of cells, carefully separating them into single cells, putting each cell into its own tiny container of nutrient. And he waited for them to reproduce, to blossom into a colony big enough to see with his naked eye. He and Milstein called those colonies of hybrid myeloma cells "hybridomas." Each hybridoma was made of identical descendants—clones—of that first single cell Köhler had separated. But were they making the antibody that they wanted? This wasn't just a random antibody, it had to be the one from the noncancerous side of the fusion, the antibody they had primed the mouse to make. The targeted antibody.

Köhler had to wait for his hybridomas to get big enough to pump out enough antibody to test. He tended them like a farmer with his seedlings—assessing their health, making sure their nutrient bath was just right and that they weren't too crowded. After weeks, when the hybridoma colonies had grown large enough and the time came for the antibody tests, Köhler was so nervous that he brought his wife with him to the basement lab to keep him calm while he looked at the results, and to buck him up if he failed.

When he saw the first results, he shouted out loud. He kissed his wife. The experiment had worked. A good number of his hybridomas were producing the antibody he'd been looking for. "That was fantastic," he said. "I was all happy."

And so it was that an Argentinian Jew and a German hippie working in a British laboratory made one of the greatest medical discoveries of the twentieth century. They did more work with these new hybridomas and the antibodies they made. What to call these antibodies to distinguish them from all others? Each hybridoma cell could be grown into rooms full of exact copies, millions of little biological factories, all chugging away day and night, making the same pure antibody. So they gave it a logical name: They were making monoclonal antibodies. They had found a way to separate and duplicate just one out of the riot of billions of different antibodies in the body, doing what the ancient alchemists had tried so hard to do: to purify a powerful single element out of the rough, wild, complex mixes of nature, making a highly targeted, natural medicine in bulk. The essential difference between monoclonal antibodies and other techniques to boost the immune system, including vaccines, was this targeted purity. Shoot a vaccine into the body, and after a lag time of a few days or weeks, the immune system responds by making scores of different types of antibodies. They can fight off a future infection. That's good. But shoot a monoclonal antibody into the body, and there is no delay. The monoclonal medicine focuses all its power against only one target, the one identified by researchers as the most vulnerable most important part of the disease process. Doctors can hit that target fast, hit it hard, and do little to upset the rest of the body. Centuries ago, Sir Thomas Brown wrote that "art is the perfection of nature." What Milstein and Köhler did was something akin to art in the laboratory. Theirs was a perfection of blood, a refinement of the body's most powerful defense system into a set of extraordinarily precise and pure medicines.

The potential of monoclonal antibodies was enormous. At the end of their first publication describing the breakthrough, Milstein and Köhler wrote, "Such cells can be grown *in vitro* in massive cultures to provide specific antibodies." And then, with wonderful understatement, "Such cultures could be valuable for medical and industrial use."

In fact, their discovery was worth a fortune.

And they didn't patent it.

THIS, TO MY MIND, is one of the most selfless and admirable moments in the history of drug discovery. It was a matter of priorities, a reflection of who Milstein and Köhler were at heart. They were true scientists, not businessmen. Their goal was to find out more about nature and to benefit humankind, not make themselves rich.

So Milstein and Köhler published their results, gave the whole game away, told the world how they did it, and in essence invited everyone to try it for themselves.

And a lot of people did. It opened a huge field of new research for other scientists. After learning the technique from Milstein and Köhler, lab after lab started making their own hybridomas, slowly building a global library of targeted antibodies. Major drug manufacturers, smelling profit, began building new labs of their own devoted to exploring this powerful new tool. It was the beginning of what we now call "biotechnology."

Milstein and Köhler became famous, of course. Prizes started being awarded, topped in 1984 with a shared Nobel for the pair (as well as for Niels Jerne, another early researcher in the field). Some of the prizes went to Milstein alone—it was his lab, after all, where the essential work had been done—and there were questions raised in the press about him getting all the credit. But the two friends didn't rise to the bait. They each remembered coming up with the idea and then convincing the other to try it. They both made important contributions to its development. One way or another, they both realized that it arose out of their friendship, and they valued that friendship over the allure of sole scientific credit. "I would not have thought about this problem in any other laboratory than César Milstein's, and I wouldn't have been encouraged to do the experiment by anyone else but him," Köhler said. Milstein, when asked, returned the compliment. When reporters prodded them, looking to stir up some controversy, they repeated variations of the same basic message: This was a joint discovery made by two friends, period.

FOR THREE YEARS after the first paper was published, both men continued working on their discovery, Milstein at Cambridge and Köhler at his next position at the Basel Institute for Immunology in Switzerland. It was getting a lot of interest as more and more immunologists learned that they could make endless amounts of targeted antibodies. Whenever anyone asked, Milstein was happy to share his techniques, his ideas, even his hybridoma cells. This was the old way of doing science. When another scientist expressed interest in following up your research, you helped them out.

It was 1978 before somebody realized there was big money to be made. That was the year researchers at the Wistar Institute in Philadelphia, one of the labs that had asked Milstein for cells, started filing patents for monoclonal antibodies they had made, directed toward viruses and cancers. Their monoclonals were made possible by Milstein's and Köhler's cells and ideas. But they had no qualms about patenting their own variations—just as drug companies do when they take another company's medicine, tweak it a little, and patent the new molecule.

Milstein was dumbstruck. He hadn't thought much about patenting. Before he and Köhler published their first paper on hybridomas, Milstein, out of courtesy to the powers that be at his institution, Cambridge, had written a note to let an official know that they had found something that might deserve patenting. But when, after a short wait, there was no reply, they went ahead and published—which meant, in Britain, that they lost most of their rights to patent. A year would pass after Milstein and Köhler's paper was published before the wheels of the British government ground out a reply to the discovery in a letter that was as clueless as it was late: "It is certainly difficult for us to identify any immediate practical applications which could be pursued as a commercial venture," the letter read.

Then the Wistar patents were filed and everyone realized that a very costly mistake had been made. There were indeed commercial possibilities tied to these cells. The Wistar patents were the start of a monoclonal gold rush. And the British were going to be frozen out.

What became known in Britain as "the patent disaster" eventually caught the attention of the prime minister herself, Margaret Thatcher, the Iron Lady. Thatcher, who had gotten a degree from Oxford in chemistry before going into politics, was outraged by the impertinence of these Wistar Americans profiting from British discoveries. It was all too reminiscent of the penicillin story, when Fleming in the 1920s had discovered the antibiotic in his London lab but couldn't purify it in bulk, so he didn't pursue it. The Americans had been the ones to figure out how to mass-produce and store it, then patented the methods and reaped the profits. Now it was all happening again. It was like a bad dream that kept repeating: a British discovery made in a British laboratory supported by British research funds was yielding exactly nothing in the way of money. Inquiries were launched. Policies were revised. Scientists were warned away from blithely sharing their ideas without going first through proper channels and ensuring that patent rights, if appropriate, were put in place. The new model for university researchers would be based around the necessity of strong patents followed by start-ups and spin-offs, commercialization and moneymaking. The old ways of open sharing and collegiality, Milstein's ways, were out.

LAB AFTER LAB, company after company started making monoclonal antibodies directed toward one target after another. This was a watershed in drug development. Instead of screening chemical after chemical in hopes that something might work on, say, a particular enzyme in a chain of reactions that led to disease—as Akira Endo had done with molds, looking for that first statin (page 211)—they could now take the target enzyme, shoot it into a mouse, create B cells that made antibodies exquisitely matched to the target, then fuse them with cancer cells to make a hybridoma that would produce monoclonal antibodies that would hit just that target. The only question was which targets were the most likely to make money.

There were technical problems, too, of course. The cells Milstein and Köhler had used in their first success were from mice, which meant that the antibodies they produced were from mice. When shot into a human, these mouse monoclonal antibodies could themselves be recognized as alien invaders—they're nonhuman, after all—setting off an immune reaction with serious side effects. Labs spent years learning to make part-mouse, part-human chimeras—the first monoclonal antibody approved by the FDA, in 1984, was about two-thirds human, one-third mouse—but the mouse sections continued to kick off immune reactions in many patients. It took years of applying the latest genetic and cell biology techniques to fully humanize the antibodies. Almost all of today's monoclonals are entirely human, and rarely kick off serious immune reactions.

The tools and techniques that had to be developed to do this "humanizing," from finding ways to switch genes off and on to using more and more accurate methods for cutting and splicing DNA and moving pieces from organism to organism, helped move other sciences ahead. The entire enterprise of working at finer and finer levels with DNA, manipulating genes as if they were puzzle pieces, ended up in triumphs like the decoding of the full human genome—and the establishment of biotechnology as the hot new seedbed of drug discovery.

Many of the new DNA techniques were immediately enlisted in the search for better ways to make all-human monoclonal antibodies. A great breakthrough was made when researchers found what's called "phage display," a clever way of enlisting bacteria and viruses to help tailor all-human antibodies.

Bio pundits began predicting that we would soon be able to identify the genes linked to diseases like cancer and Alzheimer's disease, find out what these genes were making, and then make designer monoclonal antibodies to disrupt the disease process anywhere we wanted. Monoclonal antibodies would allow us to knock out the big killers.

It hasn't turned out that way. Monoclonals have their limits. For one, they're expensive to make, requiring levels of biological expertise and high-tech equipment that cost a lot of money to bring together. They only work when they can attach to a target, which means that they only work on the surface of things. They can't get inside of cells, where a lot of the disease-causing action can take place. And they can't (yet) cross the blood-brain barrier, limiting their use for conditions there.

Even so, their use has exploded. In the early 2000s, fully human monoclonal after monoclonal began to hit the market. By 2006 they had become, as a group, the fastest-growing class of human therapeutics. In 2008 there were thirty on the market globally, and they had grown into a $30 billion industry. Six years later, there were nearly fifty being sold. The monoclonal market is expected to grow to somewhere around $140 billion by 2024.

Today's single biggest-selling drug, generating close to $20 billion per year all by itself, is Humira, a monoclonal antibody used to relieve the pain and swelling caused by some incurable autoimmune diseases including several kinds of arthritis, severe psoriasis, and Crohn's disease. It doesn't work every time—what drug does?—but it can help many patients who have no other options. It makes so much money not because of the sheer number of users, but because it's so expensive. A single shot of Humira can set patients (and their insurers) back more than $1,000. A year's treatment can cost around $50,000.

Monoclonal antibodies are the biggest thing in medicine. And it's still early days. We are now building enormous libraries of information about how antibodies are built at the atomic level, ever-more-detailed maps of their active areas, ever-finer tools to find and attack likely disease targets. Then we can create, tailor, and test a monoclonal to attack it. Monoclonals are close to perfect magic bullets.

Each new advance helps us make medicines with bigger positive effects and fewer negative ones, able to last longer in the body, effective against more diseases. They're already working well against some kinds of cancer, against inflammation in a variety of diseases, and migraines, and are showing signs of doing good things to fight Alzheimer's. In theory, the potential targets for these medicines are as numerous as the immune system is complex. We have just begun exploring the possibilities.

Costs will need to come down. Treatment with monoclonals can be very expensive, so much so that only the wealthy, those patients with very good health insurance, and the most severe cases get to benefit from them. The good news is that as more monoclonals come on line and patent protections expire, competition will increase and prices should drop. Eventually. Humira's initial patent, for instance, expired in 2016, but the company making it since 2003 has secured around one hundred additional patents covering various aspects of the drug's manufacturing processes and techniques—a wall of patents reinforced by some very highly paid lawyers. That should keep cheaper versions from showing up until around 2023.

MOST BIG PHARMA companies made their fortunes off what they call small-molecule drugs, relatively small molecules made in chemist's laboratories, then tested in much the same way Gerhard Domagk screened drugs when he discovered sulfa back in the 1920s (page 104). They got better at finding small-molecule drugs, and they got very, very good at marketing and selling them. Most of the drugs in this book are considered small molecules.

But they weren't prepared for the new age ushered in by monoclonal antibodies. Antibodies are, by comparison, huge molecules. The ways they are designed and made are rooted not in chemistry so much as the biological sciences, especially genetics and immunology. The big drugmakers had neither the mind-set nor the facilities to make the jump to biologicals. Not that they didn't try. Bayer, for instance, invested a reported half-billion dollars in a program to start making biologicals, and other big manufacturers did the same. But the old giants of drug-making were built around a different model for discovery, one that was more chemical than biological. The shift to in-house biotechnology proved too expensive in both money and time, for the most part. Besides, why build an entirely new operation when it was faster and cheaper to scan the growing number of biotech start-ups sprouting around many research universities, pick the most promising, and make a deal? You could outsource discovery.

Energized by the success of Genentech—the first major biotech company, founded in 1976 by a professor and a venture capitalist—hundreds of university researchers with bright ideas about medicine started their own spin-offs. Much of the action has now shifted to these smaller, nimbler companies. Universities began learning the art of turning the insights of their researchers into large chunks of money by hiring more lawyers and making new kinds of deals, becoming experts in protecting intellectual property, installing start-up incubators, and building research parks.

IN A WAY, this seems comforting. Universities are still repositories of great minds and breakthrough thinking, seemingly driven less by profit than by a lust for new knowledge. Seen from this angle, pure and noble science looks like it might trump the money-centered, assembly-line thinking of Big Pharma.

But viewed another way, the scene is anything but comforting. Milstein's university, Cambridge, made sure that its researchers would never again release potentially valuable research without first making sure that it was reviewed by officials, with protections put in place to assure that the institution would benefit. By now, every other major research university around the world has done the same. University scientists, fully aware that this might be the road to riches, are tailoring their work accordingly, looking for their main chance, making sure that they set up business arrangements along with scientific breakthroughs. Seen from this angle, it looks like universities and their scientists aren't fighting against the profit motive; they've been infected by it.

Of course, both ways of viewing the issue are true. It comes down to a matter of emphasis. Some researchers will be primarily motivated by the desire to ease suffering, while in others the profit motive will dominate. Both motivations are powerful, and both are valid. The hope is that, together, they'll continue to push drug discovery forward in ways that will benefit the world.

THE FUTURE OF DRUGS

IN 2003, the *British Medical Journal* breathlessly announced "the most important piece of medical news for the past fifty years." It was the coming of the polypill, that wonder drug of wonder drugs, containing within each daily pill three blood pressure drugs, a statin, folic acid, and aspirin. Its developers predicted that it could reduce heart problems by up to 80 percent and suggested that it might be used by everyone on earth over fifty-five years of age. Years of research followed. But enthusiasm waned as real-world results came in far short of hopes. The idea of the polypill is still around and still has its proponents. But not many.

A dozen years after the polypill's debut in the *BMJ*, former president Jimmy Carter announced that he was a dying man. In the summer of 2015 he had been diagnosed with an advanced case of a particularly aggressive metastatic cancer—a melanoma—that had spread to his liver and brain. His family has a history of cancer. He was in his nineties. He was declaring, in essence, his imminent death.

Then he added that his doctors were trying one last therapy, a Hail Mary, treating him with one of those new monoclonal antibodies.

Less than four months later, he issued another statement. His cancer was gone. It was not just stalled or shrinking (in remission); it had disappeared. Scans couldn't find any sign of the cancer anywhere in his body. He was cured.

This miracle was due to pembrolizumab, a monoclonal antibody drug approved by the FDA just a year earlier. It is what's called a "checkpoint inhibitor," designed to make it harder for cancers to hide from the immune system. It gave Carter's immune system a heightened ability to find and destroy the disease.

Carter was lucky—only one-quarter of cancer patients with his type of disease respond to this particular drug—but his case illustrates just how fast new drugs can change last year's death sentence into this year's survival.

Between the polypill and the president, between the prediction of wonder drugs and the reality, came years of work by pharmaceutical experts. The now-enormous global drug enterprise—Big Pharma plus all those biotech start-ups— is ceaselessly searching for the next breakthrough. What will the next round of miracles look like?

My answers are: Nobody knows, only fools would try to predict specifics, and many of the breakthroughs in drugs won't come from the old giant drug companies. Nobody knows when or even if we'll ever find a "cure" for Alzheimer's, or all cancers, or the totality of heart disease. My guess is that we will, and sooner rather than later. But that's only a guess.

What I can do with somewhat more certainty is point to some trends that will shape the world of drug research in the near-term future. Here are a few of the most important.

A shift from chemicals to biologicals

You can't have biology without chemistry, and you can't have chemical drugs without getting them to work in biological systems (like your body), so when it comes to drugs, the terms "chemicals" and "biologicals" certainly overlap. What I'm talking about here is a grand shift away from the old chemical drug discovery model—basically, "Let's test a bunch of chemicals and see if one cures a disease"—to a new paradigm that works through the manipulation of genes, cells, and microorganisms. It's more than the source of the drugs that's important here. There's also a difference in approach. Today's biotechnology companies work from a deep understanding of the disease back to the drug, doing their best to design medicines that are highly targeted to what researchers hope are weak points in the disease process. Examples range from the coming deluge of monoclonal antibodies to lab-designed replacements for damaged enzymes.

Much of the success we've had recently—like with monoclonal antibodies—is based on our new ability to manipulate DNA, the chemical instructions for our bodies—in other words, our genome. "Drug discovery is experiencing a paradigm shift," explained one expert, "whereby the explosion in genomic sciences is being harnessed to produce innovative therapies within shorter time frames."

That gets to the heart of why biologicals will be increasingly important. And it's not just *our* DNA that's in play. We're also starting to better understand and manipulate the genes of the billions of bacteria and viruses that inhabit our bodies. This hidden world inside of us, our "microbiome," helps us stay healthy in ways we're just beginning to appreciate.

Drug companies are betting heavily that these new biological approaches will pay off big time and are snapping up promising biotechnology start-ups to speed the process.

Digital drugs

Linking computers to drugs can work in a number of ways. The simplest is to put a tiny sensor into each pill that sends out a signal when the drug is taken. In early models now being tested, the sensor is about the size of a sesame seed, the power comes from chloride ions in the stomach, and the signal is picked up by a patch on the stomach. From there it can be sent to a smartphone or some other kind of transmitting device and fed into other computer systems. The first digital drug of this sort to receive FDA approval (in late 2017) was Abilify MyCite, an antipsychotic with a sensor designed to show that the drug is being swallowed on schedule. This is important for patient populations that might be inclined to miss doses, like people with mood disorders and mental illness, or the elderly, where the combination of a lot of medications plus failing memories can result in serious side effects from missed or doubled pills. If you're a conspiracy theorist, you might imagine some Big Brother future, where pills of potential pharmaceuticals of abuse, like Oxycontin and fentanyl, are enriched with nanotechnology sensors and transmitters, allowing authorities to track them wherever they go—even through someone's digestive tract.

The search for new drugs has also gone digital. Much of the action here is focused on visualizing ever-more complex drugs, all the way up to the level of huge proteins, on computers before wasting time trying to make them in a lab. Supercomputers are needed to do the calculations to show what final shape a protein is going to take after it's made, a computation challenge so extreme that we still haven't perfected it. When it happens, though, it will enable scientists to take another step in making many more highly targeted, well-tolerated drugs on their screens, in theory lowering costs and speeding the discovery process. Other computer programs can then be used to study what the newly designed protein is likely to do once it's in the body. Computer-based protein modeling is allowing drugmakers—once limited to testing *in vitro* (in the lab) and *in vivo* (in a live animal)—to do more and more *in silico*: in a computer.

The third aspect of digital drug development is less about sheer supercomputer muscle and more about communication:

using the Internet to gather information from a wider universe, and crowdsourcing some parts of drug development. The pharmaceutical company Lilly, for instance, started a website called InnoCentive, where researchers around the world are invited to come up with solutions to scientific challenges in the hope of winning prize money, with projects like finding better ways to track the behavior of individual cells, monitor viruses in wastewater, and maintain steady glucose levels in diabetes patients. Instead of trekking through rain forests looking for medicinal plants, drug researchers are now trolling the web looking for good ideas.

Here's another example: The National Institutes of Health are now recruiting subjects for what is likely to be the largest detailed health study in human history. The inelegantly named All of Us Research Program hopes to track a million-plus people, representing all of the diversity of the United States, willing to have their genomes sequenced, and then to provide access to blood test results and medical records for an indefinite time. "If all goes well," reports the *New York Times*, "the result will be a trove of health information like nothing the world has seen." This massive, big data "biobank" of information should help health professionals better understand who gets sick, and when, and why.

Another approach to crowdsourcing is being pioneered by a group of nonprofits. In 1999 a group of government and charitable organizations, worried that the pipeline for new antimalarial drugs was drying up, started the Medicines for Malaria Venture (MMV), connecting public, private, medical, government, and corporate players to find better ways to address the disease, which still kills more than a million people per year. Drug companies know that developing new antimalarials is expensive, and that the majority of potential patients are poor. Chances of profit are, therefore, small. The nonprofits wanted to develop new malaria drugs for the public good, not private gain. Could they work together?

Yes, as it turned out, they could. For example, one of their projects, launched in 2012 by the MMV along with the Bill & Melinda Gates Foundation and the drug behemoth GlaxoSmithKline, was called the Malaria Box. When requested, the MMV would send researchers a box filled with hundreds of hard-to-get experimental drugs, gathered from a variety of public and private labs, that looked like they might possibly have some benefit against malaria. The drugs were provided for free to "anyone with an interesting idea about how to use them," anywhere around the world, according to the Gates Foundation; the only thing researchers were asked to do in return was to share their results openly.

That seems a long way from digital pills. But this sort of global outreach and fast, open information-sharing is possible only thanks to computer communication. The Malaria Box model is being applied to other neglected diseases, in hopes that drug development can move out of the secretive precincts of the big drug companies and tap into what one expert calls "the global brain."

Personalized medicine

At the other end of the spectrum from the global brain lies the world of personalized medicine. With our new ability to fairly inexpensively and quickly read the details of every person's DNA—their genome—comes the opportunity to find where things have gone wrong. Each of our genes, those sections of DNA that code for an individual protein, has a small chance of being damaged in some way, with DNA missing a chunk here or getting crossed up with a bit there, or any of a number of other problems. When the DNA instructions are damaged, so are the products (the proteins the DNA codes for). Sometimes the resulting proteins might not work right, or even work at all, messing up a chain of reactions, hindering some metabolic processes, sometimes resulting in serious health problems.

Each human's genome is unique, and as a result so is each human. There's only one you. Your body has its own individual way of reacting to foods, and stresses, and sex, and everything else. It's called "biochemical and psychological individuality." Every person is different when it comes to reacting to drugs, too; the same dose can give some patients pure benefits while others get mostly side effects. No drug works exactly the same way for everybody. We are too individual. That's why, when researchers are figuring out what doses to give, they rely on statistical averages—what worked best for the greatest number of patients. There's no guarantee it will act that way for you.

Now that we can read each person's instruction manual—their DNA—we can find more of the molecular roots of that individuality and design drug regimens that are tailored just for a single person. That's the new idea of personalized medicine: Medical treatment designed to take into account our personal genetic strengths and weaknesses.

There's a lot of excited talk about the possibilities of personalized medicine, but I have trouble with the idea. I can't see everyone going in for their DNA scan and then acting on what they find out. For one thing, the line from gene to disease is rarely a straight one. The conditions we worry most about today, like Alzheimer's, cancer, and heart disease, involve not just a weakness in one gene, but the interplay of many genes over a good amount of time, plus environmental factors. That's going to take a lot more to figure out than a printout of your genes. Even when a problem with a single gene shifts the odds toward a potential health risk, there's no saying that the condition in question will actually happen to you. And if you're worried enough to do something about it, there's no guarantee that any treatment is available. Bottom line: Even after you know what's going on with your DNA, you might not be able to do anything about it—which means you're stuck for the rest of your life worrying about a molecular defect you can't fix. What's the benefit of that?

Another thing: If you're seeing a good personal physician, you're already getting a kind of personalized medicine; it's just that the personalization is done by your doctor, not a computer analysis of your DNA. It's your doctor who assesses your very personal current condition, your known health risks and habits, and comes up with a health plan tailored just for you.

Still, the vision is so enticing: A blueprint of every person's health risks, available at birth, enabling highly tailored health plans that will avoid or delay serious health issues. What could be better? So the search for reasonable applications of personalized medicine continues.

Squeezing more out of existing medicines

This isn't as sexy as some of the computer and genomics stuff, but it might just be more important: We're going to see some powerful improvements in existing drugs and treatments, along with their expansion for new uses. Part of this trend will come from advances in delivering the drugs, things like specialized pill coatings and longer-lasting formulations that don't have to be taken daily. Part will come from incremental improvements in effectiveness, as dosages and applications are refined.

This is exciting for manufacturers because they can market something new and better even though at heart it's the same drug that's already gone through all the expense of development, testing, and approval. Existing vaccines can be boosted by attaching them to new adjuvants (molecules that help wake up the immune system and improve the vaccine's effectiveness). Adding a digital sensor or developing a new time-release version can make an old drug new, creating something that can be sold to new groups of patients, expanding the market without the enormous extra cost of starting from square one.

Then there's repurposing. Once a drug's been approved for one condition, it is often found to be useful for another. So companies search for ways to reposition or repurpose their existing drugs, expanding their use for new indications. Examples include drugs like the blockbuster monoclonal antibody medication Humira, first approved in 2002 for rheumatoid arthritis, then launched in 2007 for Crohn's disease, in 2008 for psoriasis, and so on. It's now approved for no less than nine conditions, making Humira what one newspaper calls "the Swiss army knife of pharmaceutical drugs." Even that pales in comparison with the antipsychotic aripiprazole (Abilify) which has notched twenty-four approved uses.

Diseases you've never heard of

Lots of people worry about some strange new germ roaring out of the jungles of Asia or Africa, creating a pandemic that will sweep all before it.

But have you ever worried about nonalcoholic steatohepatitis (NASH)? Neither had I, until a recent article pointed out that NASH, a kind of liver disease with fatty buildup and inflammation, affects tens of millions of Americans, is linked to diabetes and obesity, and often goes undetected. In some cases it can lead to serious liver damage, and serious liver damage can kill you. You're going to hear a lot more about NASH soon, because the first of some forty drugs being tested by pharmaceutical companies are going to start hitting the market. There will be ads and news stories all over the place. Suddenly you'll worry that you or a loved one has a disease nobody had heard about a year ago. Physicians will start doing tests, patients will start to get alarmed, pills will be sold and swallowed, and enormous profits will be reaped. Some number of lives are likely to be saved, too. Then we'll all wake up to the dangerous side effects, and the Seige cycle will start all over.

Diseases you didn't know existed, like NASH—not super-dangerous but widespread, amenable to a lifetime of preventive drug treatment—will keep popping up not because they're unusually important, but because they're moneymakers. It's not that NASH isn't serious in many cases, but rather that drugs to treat it will benefit from a potentially huge market of patients who will be taking them for many years, with what are likely to be marginal benefits for most of them. It's the statins model all over again—the medicalization of our lives—at work.

Big problems for Big Pharma

The research and development model of today's pharmaceutical industry, one expert recently wrote, "is showing signs of fatigue; costs are skyrocketing, breakthrough innovation is ebbing, competition is intense, and sales growth is flattening." People inside the industry are worried that they've long since picked all the low-hanging fruit for drug development, that the complexity of testing and the amount of time needed to find new blockbuster drugs is making the whole enterprise too risky, and that in any case there are only so many targets for drugs in the body (around eight thousand potential places for drugs to work, by one estimate), so while we might not be running out of chemicals and biologicals, we're running out of bull's-eyes to aim them at. Time for a major disruption?

Perhaps so. Certainly the rise of crowdsourced research, Internet-shared data, and the increasing importance of new start-ups is making the secretive old giants of the drug world look less like nimble winners and more like lumbering dinosaurs. They will probably need to change or go extinct.

But those dinosaurs are still huge, still profitable, and have some very smart executives and researchers. They know the regulatory rules inside and out, are masters at persuading the medical community, hire effective lobbyists, and are doing their best to innovate and adapt at a rapid clip. So don't count them out.

There's another factor at play that might shift the ground from under them: Everybody hates Big Pharma. There are few fields of business more reviled by politicians, activists, and renegade researchers. The media piles on, too, when they're not helping tout the next great miracle drug.

Part of that criticism is based on the ways Big Pharma corrupts the practice of medicine. In 2002 Arthur Relman, former editor of the *New England Journal of Medicine*, bluntly sounded the alarm, writing, "The medical profession is being bought by the pharmaceutical industry, not only in terms of the practice of medicine, but also in terms of teaching and research. The academic institutions of this country are allowing themselves to be the paid agents of the pharmaceutical industry. I think it's disgraceful." Events of the past two decades, some of which I touch on in the chapter on statins (page 211), support his views.

Drug companies have mastered the art of promoting facts that support their products and minimizing or obscuring those that do not. They skew scientific findings. They court influential leading physicians, wining and dining them, hiring them as consultants and speakers. Drug representatives are experts at selling to doctors, but more recent lobbying and persuasion efforts have expanded beyond that, to ethicists, journal editors, media personalities, lawyers, politicians, patient advocacy groups, nonprofit leaders, the people who run insurance programs and oversee managed care programs, and anyone else they think might have an influence on drug sales, laws, or policies. The means are manifold, the sums are significant, and the facts have been clearly laid out in a number of recent critical books and articles.

Many physicians, and increasing numbers of politicians and the general public, are wising up. As Relman wrote, "it's disgraceful," but it's also likely to change as major manufacturers come to grips with a more vocal, more organized set of critics. What's at stake is the credibility of health care itself.

I SUDDENLY REALIZED, after writing those preceding paragraphs, that contrary to what I said in the Introduction, maybe I do have something of an agenda. If I do, it is this: to rescue the development of drugs—some of the most powerful, most beneficial medical tools ever developed—from the control of for-profit corporations. As long as Big Pharma puts money over health, they are unworthy of being the sole developers of new drugs. I think we can find other models, based on public funding for the public good.

One way or another, however, we're going to continue to reap a rich harvest from the work that's already been done. Unless society breaks apart completely, science—including drug science—will continue to move forward, gathering new knowledge, then using that knowledge to make new advances. We are likely to begin seeing great progress from collecting together all we've learned at the molecular level and using it to make major advances against our most difficult remaining disease foes: heart disease, dementia, diabetes, and cancer.

What's the future of drug development? In one line: Great things are coming.

SOURCE NOTES

I THINK IT'S a shame to burden a book like this—designed to be a relatively brisk read—with academic-style footnotes. Instead, I've pulled together the most important sources for each chapter here, so that readers who want to know more about a given drug can find more information if they want it and can see where I got my information. The author names and dates in these summaries refer to entries in the Bibliography (page 277).

INTRODUCTION

You can't really separate the history of medicine from the history of drugs. Various perspectives and approaches to understanding parts of these inter-twined histories can be found in Ban (2004), Eisenberg (2010), Gershell (2003), Greene (2007), Healy (2002, 2013), Herzberg (2009), Jones et al. (2012), Kirsch and Ogas (2017), Le Fanu (2012), just about any book from Li, Shorter (1997), Raviña (2011), Sneader (2005), Snelders (2006), Temin (1980), and Ton and Watkins (2007).

CHAPTER 1

More on the early history of opium through 1900 can be found in Bard (2000), Booth (1998), Dormandy (2006, 2012), Griffin (2004), Heydari (2013), Hodgson (2001, 2004), Holmes (2003), Kritikos and Papadaki (1967), Meldrum (2003), Musto (1991), Petrovska (2012), and Santoro (2011). For earlier perspectives on this history, see Howard-Jones (1947) and Macht (1915). See Aldrich (1994) for more on the history of women and opium addiction.

CHAPTER 2

General historical information on smallpox, Benjamin Jesty, Edward Jenner, inoculation, and vaccination in general was drawn from Razzell (1977), Pead (2003, 2017), Behbehani (1983), Institute of Medicine (2005), Rosener (2017), Jenner (1996), Hilleman (2000), Gross and Sepkowicz

(1998), Stewart and Devlin (2005), Hammarsten et al. (1979), and Marrin (2002). To find out more about Mary Wortley Montagu, one of the great overlooked heroines of medical history, see Grundy (2000, 2001), Dinc and Ulman (2007), Zaimeche et al. (2017), Aravamudan (1995), and Silverstein and Miller (1981). The tragic story of Janet Parker was constructed from contemporary news reports.

CHAPTER 3
The story of Mickey Finn and the history of chloral hydrate as the first synthetic medicine—and the first date-rape drug—was taken from Ban (2006), Inciardi (1977), Snelders et al. (2006), Jones (2011), and numerous reference and news sources. The unnerving story of the attack on Jennie Bosscheiter was mentioned in several of these sources; my version was built in great part from details provided in Krajicek (2008).

CHAPTER 4
Many of the sources listed for chapter 1, such as Booth (1998), also have information on the semisynthetics described in this chapter. In addition, I used information from Brownstein (1993), Eddy (1957), Acker (2003), Rice (2003), Payte (1991), and Courtwright (1992 and both 2015 entries), as well as contemporaneous news reports from a variety of newspapers and magazines.

CHAPTER 5
Sulfa is a fascinating and important story. My own book about its discovery (Hager, 2006) offers readers much more about Gerhard Domagk, Bayer, and the development of Prontosil, sulfanilamide, and the later sulfa drugs. Its extensive bibliography and source notes covers all the other sources used for this chapter.

CHAPTER 6
There is a bigger story to be told about the sudden appearance of mind drugs in the 1950s—not only chlorpromazine (CPZ) and the antipsychotics that followed, but tranquilizers and antidepressants as well—why they showed

up when they did, why sales were so robust, and how they've changed psychiatry, mental health care, and our attitudes toward drug-taking. Important pieces of the CPZ story and the larger context around it can be found in Alexander et al. (2011), Ayd and Blackwell (1970), Ban (2004, 2006), Baumeister (2013), Berger (1978), Burns (2006), Caldwell (1970), de Ropp (1961), Dowbiggin (2011), Eisenberg (1986, 2010), the excellent Healy (2002), Herzberg (2009), Lopez-Munoz et al. (2005), Millon (2004), Moncrieff (2009), Overholser (1956), Perrine (1996), Shorter (1997, 2011), Siegel (2005), Sneader (2002, 2005), the essential Swazey (1974), Tone (2009), Wallace and Gach (2008), and Whitaker (2002). I also drew from the many first-person accounts of the work of Henri Laborit, Jean Delay, and other early researchers of the 1950s.

INTERLUDE

Different scholars have different views of what constitutes the "golden age" of drug discovery. Some say it started back in the early 1800s with the work of researchers like Friedrich Sertürner and Justus von Liebig, chemists who began the long process of purifying, analyzing, and studying medicinal chemicals at the level of molecules. Others say it started later in that century, with Louis Pasteur's germ theory and the new focus on synthetic chemicals at companies like Bayer. But most historians focus on the three decades from 1930–1960, the period when a flood of new wonder drugs flowed from what we can now recognize as modern drug companies. This is the view held by Le Fanu (2012) and Raviña (2011), from whose work I sourced many of the facts in this short chapter.

CHAPTER 7

On the history of the Pill, see Asbell (1995), Djerassi (2009), Dhont (2010), Goldin and Katz (2002), Liao and Dollin (2012), Potts (2003), and Planned Parenthood Federation of America (2015). Much more on the Rockefeller Foundation's "Science of Man" program can be found in Kay (1993). The appearance of Viagra kicked off a storm of media coverage, some of which I used in this chapter (particularly stories in the *New York Times* and from the BBC; these can be searched online by topic), capped with the cover story in *Time* magazine (May 4, 1998) that I mention. I also relied on Campbell (2000), Goldstein (2012), and Osterloh (2015). Klotz (2005) offers an amusing first-person account of the Giles Brindley lecture.

CHAPTER 8

This was a difficult chapter to write, because it focuses on the drugs that are causing today's opioid epidemic—and demonstrates that the issues we face today are fundamentally the same as those we've faced with this family of drugs ever since the 1830s. In other words, we're making little progress in controlling our long, dysfunctional love affair with the poppy. In fact, things are getting worse. This is a hard lesson for me (a committed techno-optimist), because the nature and scale of the opioid problem is inherently pessimistic. Many of the sources I used for chapters 1 and 4 also apply here, notably Booth (1998), Acker (2003), Courtwright (both 2015 articles), and Li (2014). Additional details on Paul Janssen and fentanyl can be found in Black (2005) and Stanley (2014). And then there are the more ephemeral materials related to the current "epidemic," a wave of alarmed news items, blog entries, articles in popular magazines, and hand-wringing editorials that often gloss over facts and sometimes propose easy answers. These I used only occasionally and very selectively.

CHAPTER 9

The personal nature of my dive into statins fueled a bit of overresearching on my part. Not only did I want to get everything as right as I could when it came to my personal health, but the more I learned about statins and their marketing, the more emblematic they became of certain trends in medicine that trouble me. Because of the big money at stake and the vast numbers of people taking these drugs, there is a push and pull of controversy between statin makers and statin critics that swirls to this day. That controversy is almost as important as the medicine itself, and you'll see it reflected in many of the hundreds of papers published since the revision of guidelines earlier in this decade. Among the most important sources I used are the highly recommended Greene (2007), Agency for Healthcare Research and Quality, US DHHS (2015), Barrett et al. (2016), Berger et al. (2015), Brown and Goldstein (2004), Cholesterol Treatment Trialists' Collaborators

(2012), de Lorgeril and Rabaeus (2015), a controversial paper by Diamond and Ravnskov (2015), DuBroff and de Lorgeril (2015), Endo (2010), Fitchett et al. (2015), Garbarino (2011), Goldstein and Brown (2015), Hobbs et al. (2016), Ioannidis (2014), Julian and Pocock (2015), McDonagh (2014), Mega et al. (2015), Miller and Martin (2016), Pacific Northwest Evidence-Based Practice Center (2015), Ridker et al. (2012), Robinson and Kausik (2016), Schwartz (2011), Stossel (2008), Sugiyama et al. (2014), Sun (2014), Taylor et al. (2013), and Wanamaker et al. (2015). More details on my personal journey plus helpful tips for distinguishing good statin science from bad can be found in Hager (2016).

CHAPTER 10

Monoclonal antibodies are so new that for much of this chapter I relied on (carefully chosen) news stories and information from medical websites. The work of César Milstein and Georges Köhler is reviewed (from the perspective of Köhler's life) most fully in Eichmann (2005). The earliest full treatment of their work is found in Wade (1982). Other important sources include Yamada (2011), Buss et al. (2012), Liu (2014), Carter (2006), and Ribatti (2014). Readers who want to know more about the immune system in general might take a look at Hall (1998), which is very good, although now somewhat dated.

EPILOGUE

Speculation about the future of the pharmaceutical industry is scattered through the professional literature and the popular press. For a deeper perspective on how things are changing, see for example Gershell and Atkins (2003), Ratti and Trist (2001), Raviña (2011), Munos (2009), Hurley (2014), and Shaw (2017).

BIBLIOGRAPHY

THIS LIST INCLUDES many, but not all, of the source materials used for this book. In addition, I drew *carefully* from scores of recent newspaper and magazine articles, television transcripts, corporate reports, and web pages. I emphasize *carefully* because much of the reporting on drugs in the daily and weekly press is sensational, unbalanced, and driven by a combination of media needs to attract eyeballs and pharma/corporate needs to make money. In other words, when it comes to drugs, there's a jungle of sometimes false, often misleading, and generally overhyped claims out there in the press, on TV, and across the web (especially on social media). So let explorers beware. My research focus, as you'll see from these books and articles, tends away from the daily squawk.

Acker, Caroline Jean. "Take as Directed: The Dilemmas of Regulating Addictive Analgesics and Other Psychoactive Drugs." In *Opioids and Pain Relief: A Historical Perspective*, edited by Marcia L. Meldrum, 35–55. Seattle: IASP Press, 2003.

Agency for Healthcare Research and Quality, US Department of Health and Human Services. "Statins for Prevention of Cardiovascular Disease in Adults: Systematic Review for the U.S. Preventive Services Task Force." AHRQ Publication No. 14–05206-EF-2 (Dec. 2015).

Aldrich, Michael R. "Historical Notes on Women Addicts." *J Psychoactive Drugs* 26, no. 1 (1994): 61–64.

Alexander, G. Caleb, et al. "Increasing Off-Label Use of Antipsychotic Medications in the United States, 1995–2008." *Pharmacoepidemio. Drug Saf* 20, no. 2 (2011): 177–218.

Aravamudan, Srinivas. "Lady Mary Wortley Montagu in the *Hammam*; Masquerade, Womanliness, and Levantinization." *ELH* 62, no.1 (1995): 69–104.

Asbell, Bernard. *The Pill: A Biography of the Drug that Changed the World*. New York: Random House, 1995.

Ayd, Frank J., and Barry Blackwell. *Discoveries in Biological Psychiatry*. Philadelphia: J. B. Lippincott Co, 1970.

Ban, Thomas, et al., eds. *Reflections on Twentieth-Century Psychopharmacology*. Scotland, UK: CINP, 2004.

Ban, Thomas A. "The Role of Serendipity in Drug Discovery." *Dialogues Clin Neurosci* 8, no. 3 (2006): 335–44.

Bard, Solomon. "Tea and Opium." *J Hong Kong Branch R Asiat Soc* 40 (2000): 1–19.

Barrett, Bruce, et al. "Communicating Statin Evidence to Support Shared Decision-Making." *BMC Fam Pract* 17 (2016): 41.

Baumeister, A. "The Chlorpromazine Enigma." *J Hist Neurosci* 22, no. 1 (2013): 14–29.

Behbehani, Abbas M. "The Smallpox Story: Life and Death of an Old Disease." *Microbiol Rev* 47, no. 4 (1983): 455–509.

Berger, Philip A. "Medical Treatment of Mental Illness." *Science* 200, no. 4344 (1978): 974–81.

Berger, Samantha, et al. "Dietary Cholesterol and Cardiovascular Disease: A Systematic Review and Meta-Analysis." *Am J Clin Nutr* 102 (2015): 276–94.

Black, Sir James. "A Personal Perspective on Dr. Paul Janssen." *J Med Chem* 48 (2005): 1687–88.

Booth, Martin. *Opium: A History*. New York: St. Martin's Press, 1998.

Boylston, Arthur. "The Origins of Inoculation." *J R Soc Med* 105 (2012): 309–13.

Brown, Michael S., and Joseph L. Goldstein. "A Tribute to Akira Endo, Discoverer of a 'Penicillin' for Cholesterol." *Atheroscler Suppl* 5 (2004): 13–16.

Brown, Thomas H. "The African Connection." *JAMA* 260, no. 15 (1988): 2,247–9.

Brownstein, Michael. "A Brief History of Opiates, Opiod Peptides, and Opiod Receptors." *Proc Natl Acad Sci U S A* 90 (1993): 5,391–3.

Burns, Tom. *Psychiatry: A Very Short Introduction*. Oxford: Oxford University Press, 2006.

Buss, Nicholas, et al. "Monoclonal Antibody Therapeutics: History and Future." *Curr Opinion in Pharmacology* 12 (2012): 615–22.

Caldwell, Anne E. *Origins of Psychopharmacology: From CPZ to LSD*. Springfield, IL: Charles C. Thomas, 1970.

Campbell, S. F. "Science, Art and Drug Discovery: A Personal Perspective." *Clin Sci* 99 (2000): 255–60.

Carter, Paul J. "Potent Antibody Therapeutics by Design." *Nat Rev Immunol* 6 (2006): 343–57.

Cholesterol Treatment Trialists' Collaborators. "The Effects of Lowering LDL Cholesterol with Statin Therapy in People at Low Risk of Vascular Disease: Meta-Analysis of Individual Data from 27 Randomized Trials." *Lancet* 380 (2012): 581–90.

Courtwright, David T. "A Century of American Narcotic Policy." In *Treating Drug Problems: Volume 2: Commissioned Papers on Historical, Institutional, and Economic Contexts of Drug Treatment*, edited by Gerstein, D. R., and H. J. Harwood. Washington, D.C.: National Academies Press, 1992.

———. "The Cycles of American Drug Policy." *History Faculty Publications* 25 (2015): https://digitalcommons.unf.edu/ahis_facpub/25.

———. "Preventing and Treating Narcotic Addiction—A Century of Federal Drug Control." *NEJM* 373, no. 22 (2015): 2,095–7.

Covington, Edward C. "Opiophobia, Opiophilia, Opioagnosia." *Pain Med* 1, no. 3 (2000): 217–23.

de Lorgeril, Michel, and Mikael Rabaeus. "Beyond Confusion and Controversy, Can We Evaluate the Real Efficacy and Safety of Cholesterol-Lowering with Stains?" *JCBR* 1, no. 1 (2015): 67–92.

de Ridder, Michael. "Heroin: New Facts About an Old Myth." *J Psychoactive Drugs* 26, no. 1 (1994): 65–68.

de Ropp, Robert. *Drugs and the Mind.* New York: Grove Press, 1961.

Defalque, Ray, and Amos J. Wright. "The Early History of Methadone: Myths and Facts." *Bull Anesth Hist* 25, no. 3 (2007): 13–16.

Dhont, Marc. "History of Oral Contraception." *Eur J Contracept Reprod Health Care* 15(sup2) (2010): S12–S18.

Diamond, David M., and Uffe Ravnskov. "How Statistical Deception Created the Appearance that Statins Are Safe and Effective in Primary and Secondary Prevention of Cardiovascular Disease. *Expert Rev Clin Pharmacol* (2015): Early online, 1–10.

Dinc, Gulten, and Yesim Isil Ulman. "The Introduction of Variolation 'A La Turca' to the West by Lady Mary Montagu and Turkey's Contribution to This." *Vaccine* 25 (2007): 4,261–5.

Djerassi, Carl. "Ludwig Haberlandt—'Grandfather of the Pill.'" *Wien Klin Wochenschr* 121 (2009): 727–8.

Dormandy, Thomas. *The Worst of Evils: The Fight Against Pain.* New Haven: Yale University Press, 2006.

———. *Opium: Reality's Dark Dream.* New Haven: Yale University Press, 2012.

Dowbiggin, Ian. *The Quest for Mental Health: A Tale of Science, Scandal, Sorrow, and Mass Society.* Cambridge, UK: Cambridge University Press, 2011.

DuBroff, Robert, and Michel de Lorgeril. "Cholesterol Confusion and Statin Controversy." *World J Cardiol* 7, no. 7 (2015): 404–9.

Eddy, Nathan B. "The History of the Development of Narcotics." *Law Contemp Probl* 22, no. 1 (1957): 3–8.

Eichmann, Klaus. *Köhler's Invention.* Basel: Birkhäuser Verlag, 2005.

Eisenberg, Leon. "Mindlessness and Brainlessness in Psychiatry." *Brit J Psychiatry* 148 (1986): 497–508.

———. "Were We All Asleep at the Switch? A Personal Reminiscence of Psychiatry from 1940 to 2010." *Acta Psychiatr Scand* 122 (2010): 89–102.

Endo, Akido. "A Historical Perspective on the Discovery of Statins." *Proc Jpn Acad Ser B Phys Biol Sci* 86 (2010): 484–93.

Fitchett, David H., et al. "Statin Intolerance." *Circulation* 131 (2015): e389–e391.

Garbarino, Jeanne. "Cholesterol and Controversy: Past, Present, and Future." *Scientific American* (blog), November 15, 2011. https://blogs.scientificamerican.com/guest-blog/cholesterol-confusion-and-why-we-should-rethink-our-approach-to-statin-therapy/.

Gasperskaja, Evelina, and Vaidutis Kučinskas. "The Most Common Technologies and Tools for functional Genome Analysis." *Acta Med Litu* 24, no. 1 (2017): 1–11.

Gershell, Leland J., and Joshua H. Atkins. "A Brief History of Novel Drug Technologies." *Nat Rev Drug Discov* 2 (2003): 321–7.

Goldin, Claudia, and Lawrence F. Katz. "The Power of the Pill: Oral Contraceptives and Women's Career and Marriage Decisions." *J Polit Econ* 110, no. 4 (2002): 730–70.

Goldstein, Irwin. "The Hour Lecture That Changed Sexual Medicine—the Giles Brindley Injection Story." *J Sex Med* 9, no. 2 (2012): 337–42.

Goldstein, Joseph L., and Michael S. Brown. "A Century of Cholesterol and Coronaries: From Plaques to Genes to Statins." *Cell* 161 (2015): 161–72.

Greene, Jeremy A. *Prescribing by Numbers: Drugs and the Definition of Disease.* Baltimore: Johns Hopkins University Press, 2007.

Griffin, J. P. "Venetian Treacle and the Foundation of Medicines Regulation." *Brit J Clin Pharmacol* 58, no. 3 (2004): 317–25.

Gross, Cary P., and Kent A. Sepkowicz. "The Myth of the Medical Breakthrough: Smallpox, Vaccination, and Jenner Reconsidered." *Int J Inf Dis* 3, no. 1 (1998): 54–60.

Grundy, Isobel. "Montagu's Variolation." *Endeavour* 24, no. 1 (2000): 4–7.

———. *Lady Mary Montagu: Comet of the Enlightenment.* Oxford, UK: Oxford University Press, 2001.

Hager, Thomas. *The Demon Under the Microscope.* New York: Harmony Books, 2006.

———. *Understanding Statins.* Eugene, OR: Monroe Press, 2016.

Hall, Stephen S. *A Commotion in the Blood: Life, Death, and the Immune System.* New York: Henry Holt and Company, 1998.

Hammarsten, James F., et al. "Who Discovered Smallpox Vaccination? Edward Jenner or Benjamin Jesty?" *Trans Am Clin Climatol Assoc* 90 (1979): 44–55.

Healy, David. *The Creation of Psychopharmacology.* Cambridge, MA: Harvard University Press, 2002.

———. *Pharmageddon.* Berkeley: University of California Press, 2013.

Herbert, Eugenia. "Smallpox Inoculation in Africa." *J Afr Hist* XVI(4) (1975): 539–59.

Herzberg, David. *Happy Pills in America: From Miltown to Prozac.* Baltimore: Johns Hopkins University Press, 2009.

Heydari, Mojtaba, et al. "Medicinal Aspects of Opium as Described in Avicenna's *Canon of Medicine*." *Acta Med Hist Adriat* 11, no. 1 (2013): 101–12.

Hilleman, Maurice R. "Vaccines in Historic Evolution and Perspective: A Narrative of Vaccine Discoveries." *Vaccine* 18 (2000): 1,436–47.

Hobbs, F. D. Richard, et al. "Is Statin-Modified Reduction in Lipids the Most Important Preventive Therapy for Cardiovascular Disease? A Pro/Con Debate." *BMC Med* 14 (2016): 4.

Hodgson, Barbara. *In the Arms of Morpheus*. Buffalo, NY: Firefly Books, 2001.

———. *Opium: A Portrait of the Heavenly Demon*. Vancouver: Greystone Books, 2004.

Holmes, Martha Stoddard. " 'The Grandest Badge of His Art': Three Victorian Doctors, Pain Relief, and the Art of Medicine." In *Opioids and Pain Relief: A Historical Perspective*, edited by Marcia L. Meldrum, 21–34. Seattle: IASP Press, 2003.

Honigsbaum, Mark. "Antibiotic Antagonist: The Curious Career of René Dubos." *Lancet* 387, no. 10014 (2016): 118–9.

Howard-Jones, Norman. "A Critical Study of the Origins and Early Development of Hypodermic Medication." *J Hist Med Allied Sci* 2, no. 2 (1947): 201–49.

Hurley, Dan. "Why Are So Few Blockbuster Drugs Invented Today?" *New York Times Magazine*, November 13, 2014.

Inciardi, James A. "The Changing Life of Mickey Finn: Some Notes on Chloral Hydrate Down Through the Ages." *J Pop Cult* 11, no. 3 (1977): 591–6.

Institute of Medicine, Board on Health Promotion and Disease Prevention, Committee on Smallpox Vaccination Program Implementation. *The Smallpox Vaccination Program: Public Health in an Age of Terrorism*. Washington, D.C.: National Academies Press, 2005.

Ioannidis, John P. "More Than a Billion People Taking Statins? Potential Implications of the New Cardiovascular Guidelines." *JAMA* 311, no. 5 (2014): 463.

Jenner, Edward. *Vaccination Against Smallpox*. Amherst, MA: Prometheus Books, 1996.

Jones, Alan Wayne. "Early Drug Discovery and the Rise of Pharmaceutical Chemistry." *Drug Test Anal* 3 (2011): 337–44.

Jones, David S., et al. "The Burden of Disease and the Changing Task of Medicine." *NEJM* 366, no. 25 (2012): 2,333–8.

Julian, Desmond G., and Stuart J. Pocock. "Effects of Long-Term Use of Cardiovascular Drugs." *Lancet* 385 (2015): 325.

Kay, Lily. *The Molecular Vision of Life: Caltech, The Rockefeller Foundation, and the Rise of the New Biology*. New York: Oxford University Press, 1993.

Kirsch, Donald R., and Ogi Ogas. *The Drug Hunters*. New York: Arcade Publishing, 2017.

Klotz, L. "How (Not) to Communicate New Scientific Information: A Memoir of the Famous Brindley Lecture." *BJU Int* 96, no. 7 (2005): 956–7.

Krajicek, David J. "The Justice Story: Attacked by the Gang." New York *Daily News*, October 25, 2008.

Kritikos, P. G., and S. P. Papadaki. "The History of the Poppy and of Opium and Their Expansion in Antiquity in the Eastern Mediterranean Area." United Nations Office on Drugs and Crime (1967). http://www.unodc.org/unodc/en /data-and-analysis/bulletin/bulletin_1967–01–01_3_page004.html.

Le Fanu, James. *The Rise and Fall of Modern Medicine* (Revised Ed.). New York: Basic Books, 2012.

Li, Jie Jack. *Laughing Gas, Viagra, and Lipitor: The Human Stories Behind the Drugs We Use.* Oxford, UK: Oxford University Press, 2006.

———. *Blockbuster Drugs.* Oxford, UK: Oxford University Press, 2014.

Liao, Pamela Verma, and Janet Dollin. "Half a Century of the Oral Contraceptive Pill." *Can Fam Physician* 58 (2012): e757–e760.

Liu, Justin K. H. "The History of Monoclonal Antibody Development—Progress, Remaining Challenges and Future Innovations." *Ann Med Surg* 3 (2014): 113–6.

Lopez-Munoz, Francisco, et al. "History of the Discovery and Clinical Introduction of Chlorpromazine." *Ann Clin Psychiatry* 17, no. 3 (2005): 113–35.

Macht, David I. "The History of Opium and Some of Its Preparations and Alkaloids." *JAMA* 64, no. 6 (1915): 477–81.

Magura, Stephan, and Andrew Rosenblum. "Leaving Methadone Treatment: Lessons Learned, Lessons Forgotten, Lessons Ignored." *Mt Sinai J Med* 68, no. 1 (2001): 62–74.

Majno, Guido. *The Healing Hand.* Cambridge: Harvard University Press, 1975.

Marrin, Albert. *Dr. Jenner and the Speckled Monster.* New York: Dutton Children's Books, 2002.

McDonagh, Jonathan. "Statin-Related Cognitive Impairment in the Real World: You'll Live Longer, but You Might Not Like It." *JAMA Intern Med* 174, no. 12 (2014): 1,889.

Mega, Jessica L., et al. "Genetic risk, Coronary Heart Disease Events, and the Clinical Benefit of Statin Therapy: An Analysis of Primary and Secondary Prevention Trials." *Lancet* 385, no. 9984 (2015): 2,264–71.

Meldrum, Marcia L., ed. *Opioids and Pain Relief: A Historical Perspective.* Seattle: IASP Press, 2003.

Miller, P. Elliott, and Seth S. Martin. "Approach to Statin Use in 2016: An Update." *Curr Atheroscler Rep* 18 (2016): 20.

Millon, Theodore. *Masters of the Mind: Exploring the Story of Mental Illness from Ancient Times to the New Millennium.* New York: John Wiley & Sons, 2004.

Moncrieff, Joanna. *The Myth of the Chemical Cure: A Critique of Psychiatric Drug Treatment.* New York: Palgrave Macmillan, 2009.

Munos, Bernard. "Lessons from 60 years of Pharmaceutical Innovation." *Nat Rev Drug Discov* 8 (2009): 959–68.

Musto, David F. "Opium, Cocaine and Marijuana in American History." *Scientific American* (July 1991): 20–27.

Osterloh, Ian. "How I discovered Viagra." *Cosmos Magazine*, April 27, 2015.

Overholser, Winfred. "Has Chlorpromazine Inaugurated a New Era in Mental Hospitals?" *J Clin Exp Psychophathol Q Rev Psychiatry Neurol* 17, no. 2 (1956): 197–201.

Pacific Northwest Evidence-Based Practice Center. "Statins for Prevention of Cardiovascular Disease in Adults: Systematic Review for the U.S. Preventive Services Task Force." *Evidence Synthesis* 139 (2015).

Payte, J. Thomas. "A Brief History of Methadone in the Treatment of Opioid Dependence: A Personal Perspective." *J Psychoactive Drugs* 23, no. 2 (1991): 103–7.

Pead, Patrick J. "Benjamin Jesty: New Light in the Dawn of Vaccination." *Lancet* 362 (2003): 2,104–9.

———. *The Homespun Origins of Vaccination: A Brief History.* Sussex: Timefile Books, 2017.

Perrine, Daniel M. *The Chemistry of Mind-Altering Drugs: History, Pharmacology, and Cultural Context.* Washington, D.C.: American Chemical Society, 1996.

Petrovska, Biljana Bauer. "Historical Review of Medicinal Plants' Usage." *Pharmacogn Rev* 6, no. 11 (2012): 1–5.

Planned Parenthood Federation of America. *The Birth Control Pill: A History.* 2015. https://www.plannedparenthood.org/files/1514/3518/7100/Pill_History_FactSheet.pdf

Pringle, Peter. *Experiment Eleven.* New York: Walker & Company, 2012.

Potts, Malcolm. "Two Pills, Two Paths: A Tale of Gender Bias." *Endeavour* 27, no. 3 (2003): 127–30.

Ratti, Emiliangel, and David Trist. "Continuing Evolution of the Drug Discovery Process in the Pharmaceutical Industry." *Pure Appl Chem* 73, no. 1 (2001): 67–75.

Raviña, Enrique. *The Evolution of Drug Discovery: From Traditional Medicines to Modern Drugs.* Weinheim, Germany: Wiley-VCH, 2011.

Razzell, Peter. *The Conquest of Smallpox.* Sussex, UK: Caliban Books, 1977.

Ribatti, Domenico. "From the Discovery of Monoclonal Antibodies to Their Therapeutic Application: An Historical Reappraisal." *Immunol Lett* 161 (2014): 96–99.

Rice, Kenner C. "Analgesic Research at the National Institutes of Health: State of the Art 1930s to Present." In *Opioids and Pain Relief: A Historical Perspective*, edited by Marcia L. Meldrum, 57–83. Seattle: IASP Press, 2003.

Ridker, Paul M., et al. "Cardiovascular Benefits and Diabetes Risks of Statin Therapy in Primary Prevention: An Analysis from the JUPITER Trial." *Lancet* 380, no. 9841 (2012): 565–71.

Robins, Nick. "The Corporation That Changed the World: How the East India Company Shaped the Modern Multinational." *Asian Aff* 43, no. 1 (2012): 12–26.

Robinson, Jennifer G., and Ray Kausik. "Moving Toward the Next Paradigm for Cardiovascular Prevention." *Circulation* 133 (2016): 1,533–6.

Rosner, Lisa. *Vaccination and Its Critics.* Santa Barbara: Greenwood, 2017.

Santoro, Domenica, et al. "Development of the concept of pain in history." *J Nephrol* 24(S17) (2011): S133–S136.

Schwartz, J. Stanford. "Primary Prevention of Coronary Heart Disease with Statins: It's Not About the Money." *Circulation* 124 (2011): 130–2.

Shaw, Daniel L. "Is Open Science the Future of Drug Development?" *Yale J Bio Med* 90 (2017): 147–51.

Shorter, Edward. *A History of Psychiatry: From the Era of the Asylum to the Age of Prozac.* New York: John Wiley & Sons, 1997.

Shorter, Edwin, ed. *An Oral History of Neuropsychopharmacology, The First Fifty Years, Peer Interviews,* vol. 1. Brentwood, TN: ACNP, 2011.

Siegel, Ronald K. *Intoxication: The Universal Drive for Mind-Altering Drugs.* Rochester: Park St. Press., 2005

Silverstein, Arthur M., and Genevieve Miller. "The Royal Experiment on Immunity: 1721–22." *Cellular Immunol* 61 (1981): 437–47.

Sneader, Walter. "The 50th Anniversary of Chlorpromazine." *Drug News Perspect* 15, no. 7 (2002): 466–71.

———. *Drug Discovery: A History.* Sussex, UK: John Wiley & Sons, 2005.

Snelders, Stephen, et al. "On Cannabis, Chloral Hydrate, and the Career Cycles of Psychotropic Drugs in Medicine." *Bull Hist Med* 80 (2006): 95–114.

Stanley, Theodore H. "The Fentanyl Story." *J Pain* 15, no. 12 (2014): 1,215–26.

Stewart, Alexandra J., and Phillip M. Devlin. "The History of the Smallpox Vaccine." *Journal of Infect* 52 (2005): 329–34.

Stossel, Thomas P. "The Discovery of Statins." *Cell* 134 (2008): 903–5.

Sugiyama, Takehiro, et al. "Different Time Trends of Caloric and Fat Intake Between Statin Users and Nonusers Among US Adults: Gluttony in the Time of Statins?" *JAMA Intern Med* 174, no. 7 (2014): 1,038–45.

Sun, Gordon H. "Statins: The Good, the Bad, and the Unknown." *Medscape,* October 10, 2014.

Swazey, Judith P. *Chlorpromazine in Psychiatry: A Study of Therapeutic Innovation.* Cambridge, MA: MIT Press, 1974.

Taylor, Fiona, et al. "Statin Therapy for Primary Prevention of Cardiovascular Disease." *JAMA* 310, no. 22 (2013): 2,451–2.

Temin, Peter. *Taking Your Medicine: Drug Regulation in the United States.* Cambridge: Harvard University Press, 1980.

Tone, Andrea. *The Age of Anxiety.* New York: Basic Books, 2009.

Tone, Andrea, and Elizabeth Siegel Watkins. *Medicating Modern America: Prescription Drugs in History.* New York: New York University Press, 2007.

Wade, Nicholas. "Hybridomas: The Making of a Revolution." *Science* 215, no. 26 (1982): 1,073–5.

Wallace, Edwin R., and John Gach, eds. *History of Psychiatry and Medical Psychology.* New York: Springer, 2008.

Wanamaker, Brett L., et al. "Cholesterol, Statins, and Dementia: What the Cardiologist Should Know." *Clin Cardiol* 38, no. 4 (2015): 243–50.

Whitaker, Robert. *Mad in America: Bad Science, Bad Medicine, and the Enduring Mistreatment of the Mentally Ill.* New York: Basic Books, 2002.

Yamada, Taketo. "Therapeutic Monoclonal Antibodies." *Keio J Med* 60, no. 2 (2011): 37–46.

Zaimeche, Salah, et al. "Lady Montagu and the Introduction of Smallpox Inoculation to England." www.muslimheritage.com/article/lady-montagu-and-introduction-smallpox-inoculation-england.

INDEX

abscess, 112–13
abuse
    deterrents from, 207
    opioids and, 7, 98
    of patients, 136
ACC. *See* American College of Cardiology
accidents
    deaths and, 79
    discovery by, 177
    overdoses and, 38
addiction, 32, 94
    addicts and, 27
    in China, 95
    concept of, 45
    danger and, 43
    epidemic, 86
    first description of, 18
    opioids and, 85
    opium and, 20, 39
    soldiers and, 33
    terminology and, 91–92
    tobacco and, 30
    in United States, 203–5
    veterans and, 198
addicts, 46
adrenaline, 127
advertisement, 90, *93*
    birth control and, *169*
    penicillin in, *119*
    Thorazine in, *147*
advice, 211
Agawam Hunt Club, 111
AHA. *See* American Heart Association
alchemy, 19, 29
alcohol, 92
    alcoholism and, 38
alkaloids, 40–41, 96
American Civil War, 43
American College of Cardiology (ACC),
    229
American Heart Association (AHA),
    229–30
the American Medical Association, 91,
    115

amidon, 190
amputation, 105
anecdotes, 28, 43
anesthetics, 23
Animal Chemistry (Liebig), 78
animals, 12, 106
    antibiotics and, 122
    experimentation with, 25
    infection and, 66
    testing on, 79, 87
antibiotic resistance, 121–22
antibiotics, 100, 114
    age of, 119
    vaccines and, 5
    viruses and, 121
antibodies
    immune system and, 244
    monoclonal antibodies, 241–67
antihistamines, 130–31, 146, 160
anti-inoculation movement, 62–63
antipsychotics, 8, 82
    psychiatry and, 156
    trade names of, 157
anxiety, 128, 209
Archaeology, 12
arthritis, 185
aspirin, 8, 91
asylums, 133–56
ataraxia, 128–29
atoms, 83
    addition of, 86
attitudes, 4, 18
autoimmune disease, 243, 254
Avicenna (Ibn Sina), 18
    with pupils, 19, *19*
azo dyes, 107–10
    sulfanilamide and, 109

bacteria, 99–100
bans, 32, 36
The Bayer company, 87–92, *89*
    Domagk at, 105
    laboratories at, 101
    patents and, 110

Bedson, Henry, 70–72
benefits, 48
  risk analysis and, 73
Big Pharma, 83
  issues and, 268–70
  profitability and, 255–56
biologicals, 260–61
biology
  mental illness and, 138
  molecular biology, 112
birth control pill, 166–73
bitterness, 11
  alkaloids and, 40
blood, 241–57
Boilly, L. L., 65
books, 21
Bosschieter, Jennie, 80–81
Boswell, James, 25
botany, 170
brain, 201
  mental illness and, 156
  studies of, 156
Brindley, Giles, 175–77
British Medical Journal, 259
the British Museum, 1
Brompton cocktail, 96
Brown, Thomas, 248

cabeza de negro, 170–71
Calvert Lithographing Co., *93*
cannabis, 14, 92
care, 71
  trends in, 206
Caroline of Ansbach, 59–60, *60*
Carter, Jimmy, 259–60
Celebrex, 185
celebrities, 111
  endorsements by, 62
  in the White House, 199
Celestial Empire, 30
cells, 160, 200, 252
  growth of, 249
Centers for Disease Control,
  U.S., 72
chemicals, 88
  biologicals and, 260–61
  du Pont family, 111
chemistry, 75
children
  deaths of, 115
  engrafting on, 57

China, 205
  addiction in, 95
  chi flow and, 77
  Communist government in, 36
  indentured workers from,
    34–35
  opium in, 29–32
chloral hydrate (knockout drops), 8
  Big Pharma and, 84
  emergence of, 78–81
chloroform, 78–79
chlorpromazine (CPZ), 145–57
  prescriptions for, 152
  strait jackets and, 150, 155
cholera, 26
cholesterol, 214–29
Christianity, 76
cities, 96
clinical chemistry, 78
clinical trials, 61
coal tar, 87
  dyes from, 103
cocaine, 92
codeine, 41
cognition, 234–35
Collier's magazine, 114
Communism, 36
Comstock Laws, U.S.,
  168–69
constipation, 40
contagiousness, 52
  disease and, 68, 71
controversy, 230
Coolidge, Calvin, 100
coolie trade, 34–35
cough syrup, 38, 41
  heroin in, *89, 90*
cowpox, 66–68
CPZ. *See* chlorpromazine
credibility, 270
crematorium, 72
Crete, 12
crime, 94
culture shock, 213
cures, 37–38

dairymaids, 65–66
dangers, 48
Daoguang, 34
data, 212
date-rape, 81

deaths, 34, 72
  accidents and, 79
  association with, 17–18
  causes of, 4–5
  of children, 115
  mice, 106
  patients and, 127
  Pierrepont after, 65
  rate of, 120
  side effects and, 173
  small pox, 50
  Wanrong, 36
Delay, Jean, 138–46
De Materia Medica (Dioscorides), 14
demonstrations, 61
  of inoculation, 59
  for nobility, 62
Deniker, Pierre, 141–46
Department of Agriculture, U.S., 116
depression, 25, 209
diabetes, 233
diethylene glycol, 115
digital drugs, 261–64
Dioscorides, Pedanius, 14
  on sap collection, 16
diseases, 5
  bacteria and, 99
  causes of, 51
  contagiousness and, 68, 71
  new forms of, 267
  patterns in, 26
  understanding, 52–53
doctors, 3. *See also* physicians
Domagk, Gerhard, 104–11, *107*
  Nobel Prize for, 117
  sulfa and, 117, 255
dosage, 40
  accuracy and, 42
  prescriptions and, 1–2
  side effects and, 154
drowning, 125
drug rehabilitation, 45
drugs, 6, 75
  bans on, 32, 36
  discovery of, 120
  future of, 259–70
  golden age of, 159–62
  guides to, 14
  hypnotics class of, 79
  life quality and, 186
  as magic bullets, 99–122

medicine and, 2
  before nineteenth century, 16
  politics and, 151
  receptors and, 207
  sex and, 163–86
  side effects and, 230–31
  the sixties and, 192, 198
  standardization of, 24
  studies for, 23
  subculture and, 35
  synthetic, 82
du Pont, Ethel, 111, 113

The East India Company, 28, 32
economics, 209
ECT. *See* electroconvulsive therapy
ED. *See* erectile dysfunction
education, 49
effectiveness, 7
  decline in, 108
  demonstration of, 59
  gauges on, 87
  opium and, 17
  for patients, 42
Ehrlich, Paul, 101–4, *102*
elderly
  mental illness in, 152
  overweight and, 213
electroconvulsive therapy (ECT), 140
elites, 36
Endo, Akira, 213–16, 252
endorphins, 201
England, 50
  tea in, 30
English Channel, 124
engrafting, 55–58
enkephalin, 201
epidemics, 7
  addiction, 86
  bacteria and, 99–100
  morphinism as, 44
  small pox, 50–51, 58
eradication, 68, 74
erectile dysfunction (ED), 175, 179–83
estrogen, 172
ether, 92
euphoria, 17
Europe
  explorers from, 52
  inoculation in, 65
  opium in, 20, 23, 26

exhibitions, 1–9
experimentation, 61
  animals and, 25
  Jesty and, 67
  Paracelsus and, 22
  safety and, 56

failures, 86
  azo dyes and, 108
  Ehrlich and, 102, *102*
FDA. *See* Food and Drug Administration
FDR Jr. *See* Roosevelt, Franklin Delano, Jr.
fear, 127
Federal Food, Drug, and Cosmetic Act, U.S.
  (1938), 116
fentanyl, 98
fertilizers, 78
fever, 56
Fewster, John, 66
Finn, Michael, 81–82
fleet, 31, *31*
  pirate junks, 34
Fleming, Alexander, 118
Food and Drug Administration (FDA), 115
Ford, Henry, 103
foreign ambassadors, 53
Fortune magazine, 148
the four humors, 58
  theory of, 63
France, 124
Frankenstein (Shelley, M.), 77
friendship, 55
Fu Manchu, 95

Ganges River, *31*
gas gangrene, 105
genocide, 52
genome, 73
Germany, 101
germ theory, 69
the Gettysburg Address, 67
Godwin, Mary Wollstonecraft, 27
Goldin, Claudia, 174
government, 94
  control and, 207
  funding from, 152
  regulation and, 47
  support by, 64
government communism, 36
Greece, 13
Greene, Jeremy A., 222

gunpowder, 111
gynecologists, 170

Hamlin's Wizard Oil, *93*
Hanfstaengl, F., 77
happiness, 17
the Harrison Act (1914),
  93–94
  reinterpretation, 95
Harvey, Gideon, 25
healing
  art of, 22
  healers and, 76–77
health history, 206
heart attacks, 173
heart disease, 211–39
herbs, 3, 29
  medicine from, 13–14
herd immunity, 74
heroin, 85–98
  in cough syrup, 89, 90
  methadone and, 191–95
  prescriptions for, 96
heroisch (heroic), 89
historians, 32, 67
  on approaches, 129
  on golden age, 159–62
  on sleeping pill, 79
history, 9, 95
  drugs and, 2, 6
  health and, 206
  medicine and, 2–3
Hitler, Adolf, 109, 112
Hoechst labs, 187–90
Hoffmann, Felix, 88
homelessness, 153
Homer, 13
Hong Kong, 33
hormones, 167
Huguenard, Pierre, 130–31
hul gil (joy plant), 14
human body, 55
  immune system of, 69, 242
  temperature of, 125
  understanding of, 6, 112
hunter-gatherers, 11
hydrocodone, 97
hypothermia, 125

Ibn Sina. *See* Avicenna
the Iliad (Homer), 13

immune system, 248
 antibodies and, 244
 of human body, 69, 242
indentured workers, 34–35
India, 28
 opium plantations in, 30
infections
 animals and, 66
 wounds and, 105, 117
information, 206
informed consent, 61, 172
 Pincus and, 173
ingredients
 in laudanum, 22
 medicine and, 13
 opium as, 16–17
innovation, 207
inoculation, 56. *See also* vaccines
 anti-inoculation movement and, 62–63
 demonstration of, 59
 in Europe, 65
 practice of, 64
 prisoners and, 60–61
International Harvester, 168
International Opium Conference, 93
in vitro, 249

Jefferson, Thomas, 37
Jenner, Edward, 64, 65, 67
Jerne, Niels, 249
Jesty, Benjamin, 66–68
Johns Hopkins, 113
Jones, John, 25
the joy plant (*hul gil*). *See* opioids

Katz, Lawrence, 174
Klarer, Josef, 107–10, 120
knowledge, 21
Köhler, Georges, 245–52
 Nobel Prize for Milstein and, 249

laboratories, 252
 The Bayer company, 101
Laborit, Henri, 123–45
La Brune, 37
the Lasker Award, 155–56
laudanum, 22, 46
 overdoses on, 27
 by Sydenham, 24–25
 variations on, 37
laughing gas (nitrous oxide), 92

Le Fanu, James, 159
legislation, 120
Leibniz, Wilhelm Gottfried, 59
letters, 117
library, 6
 private, 49
von Liebig, Justus, 77, 77–79
life, definition of, 76
life, quality of, 186
life expectancy, 4
 in United States, 120
Life magazine, 119
life span
 antibiotics and, 100
 overdoses and, 98
Lincoln, Abraham, 67
literature, 127
 travel literature, 54
lithographs, 31, 54
lobotomies, 156
logical bassoon, 175
London Times, 35

Maalin, Ali Maow, 70
magic balls (*Zauberkugeln*), 101
Maitland, Charles, 57
malaria, 27, 69
The March of Medicine (TV show),
 146
Marker, Russell, 170
Mather, Cotton, 64
May Revolution, 155
McCormick, Katharine, 166–70, 169
Medicare/Medicaid programs, 152
medicine, 99
 art of, 20
 breakthrough in, 212
 chemistry and, 75
 drugs and, 2
 dye based, 102
 experiments and, 61
 from herbs, 13–14
 history and, 2–3
 ideas in, 21
 ingredients in, 13
 opium in, 38
 patent medicine, 37, 93, 94
 personalization of, 264–66
 practice of, 4, 57–58
 science as, 62
 success in, 53

men, 175
mental health hospitals, 82, 151
  wealth and, 153
mental illness, 132–46
  biology and, 138
  brain and, 156
  elderly and, 152
Merck pharmaceuticals, 41
metabolisms, 129
methadone, 190–95
  heroin and, 191–95
  programs and, 198
mice, 106
  streptococci in, 108
The Mickey Finn, 75–83
the Middle Ages, 19
the Middle East, 11
military, 33
  vaccination and, 73
Milstein, César, 243–52
  Nobel Prize for Köhler and, 249
Min Chueh Chang, 167
mind drugs, 148–49
mineral elements, 78
Minoans, 12
mold, 118
  penicillin as, 213
molecular biology, 112
molecules, 75, 83
  alteration of, 88
  azo dyes, 107–10
  carbon, 86
  study of, 97
  urea, 77
monoclonal antibodies, 241–67
Montagu, Edward Wortley, 50
Montagu, Mary Wortley. See Pierrepont, Mary
morphine, 40
  the Pravaz and, 43
  United States and, 85
  women and, 44–45
murder, 47, 81
muscles, 231–33
Muslims, 54–55
myths, 13

naloxone, 202–3
narcan, 203
nature, 21
nausea, 146
neurotransmitters, 156

Newgate Prison, 60–61
New York Times, 114
Nightingale, Florence, 28
nitrous oxide (laughing gas), 92
the Nobel Prize, 101
  for Domagk, 117
  for Milstein and Köhler, 249
nobility, 49
  demonstrations for, 62
  women as, 55

the Odyssey (Homer), 13
operations, 126–27
opiates, 41
  per capita use of, 43
  semisynthetic, 97
opioids, 206
  abuse of, 7, 98
  addiction and, 85
  opium and, 8–9
opium
  addiction and, 20, 39
  China and, 29–32
  in Europe, 20, 23, 26
  the joy plant as, 11–48
  in medicine, 38
  opioids and, 8–9
  seizures of, 33
opium dens, 35, 35
The Opium Wars, 32–33
organic chemistry, 41
  field of, 76
orphans, 64
the Ottoman Empire, 53
overdoses, 4, 47
  accidents and, 38
  on chloral hydrate, 80
  laudanum and, 27
  lifespan and, 98
  opioids and, 204
  reports of, 91
  rise in, 208
  risk of, 18
Oxycontin, 97–98

pain, 208
  muscles, 231–33
Papaver somniferum (opium poppy), 15
Paracelsus, 20–23, 21
parasites, 69, 121
Parker, Janet, 70–72

Passaic River, 80
patents, 88, 249
    The Bayer company and, 110
    expiration of, 255
    medicine and, 37, 93, 94
    protection of, 90, 222
patients, 24, 29
    abuse of, 136
    deaths and, 127
    effectiveness for, 42
    happiness for, 17
    operations on, 126–27
    quarantine for, 70, 71
    small pox for, 51
    tuberculosis and, 89
penicillin, 8
    in advertisement, 119
    availability of, 101
    as mold, 213
penis, 176
Pfizer, 177–85
pharmacology, 23
physicians, 18
    crime and, 95
    licenses and, 116
    in the Middle Ages, 19
    procedures for, 56–58
    profitability for, 63–64
    tools for, 17, 99
    treatments and, 51
Pierrepont, Mary, 49–74, 54
    achievement of, 65
    scars on, 53
pills
    exhibition of, 1–9
    laudanum as, 22
Pincus, Gregory, 167–73
    informed consent and, 173
    Rock and, 172
plants
    power of, 12
    species of, 13, 15–16
pleasure, 25
pneumonia, 106
Poe, Edgar Allan, 38
poetry, 50
poison, 80
polio, 70, 74
politics, 151
    the Republican Party and,
        199

poppies, 12
    American Civil War and, 43
    De Materia Medica on, 14
    Papaver somniferum, 15
    species of, 15–16
population, 85
    displacement of, 34
    protection of, 68
post-traumatic stress disorder (PTSD), 126
    veterans and, 157
power, 6
    of plants, 12
Pravaz, Charles Gabriel, 42–43
the Pravaz, 42–43, 44
pregnancy, 174
preparations, 40
Prescribing by Numbers (Greene), 222
prescriptions
    antibiotics and, 122
    black market for, 207
    for CPZ, 152
    dosage of, 1–2
    for heroin, 96
Presley, Elvis, 199
prices, 2
prisoners, 153
    inoculation and, 60–61
privacy, 39
procedures
    adoption of, 62
    demonstration of, 59
    engrafting as, 56–58
    improvement on, 63
processing, 16
profitability
    Big Pharma and, 255–56
    research and, 149
    revenue and, 63–64
Prontosil, 110–14
prostitution, 81, 92
    laudanum and, 26
psychiatry, 150
psychopharmocology, 150
PTSD. See post-traumatic stress disorder
publications, 6
Pure Food and Drug Act (1906), 93

rabbits, 106
rape, 80, 81
reactants, 76
Reagan, Nancy, 199

receptors, 199–203
  drugs and, 207
regulation, 3, 23
  government and, 47
  morphine and, 86
Reid, Wallace, 96
remedies, 22
  La Brune as, 37
  over-the-counter, 46, 91
reports, 111
  overdoses and, 91
the Republican Party, 199
research, 41
  informed consent and, 173
  profitability and, 149
researchers, 39
Resnais, Alain, 156
results, 206
revenue, 44
  practices and, 58
  profitability and, 63–64
Rhône-Poulenc (RP), 131, 140, 145
  rights sale by, 146
risks, 157, 222
  overdose as, 18
  risk/benefit analysis, 73
Rock, John, 170
  Pincus and, 172
the Romantic Era, 26
Roosevelt, Eleanor, 112
Roosevelt, Franklin Delano, Jr. (FDR Jr.),
  111–14
Roosevelt, Theodore (TR), 92
the Royal Society, 49
RP. See Rhône-Poulenc
RP-4560, 129–45

Sacher, Eberhard, 46–47
safety
  experimentation and, 56
  proof of, 60
  standards for, 71
Salvarsan, 102–3
Sanger, Margaret, 167–70, 168
scars, 51–52
  Pierrepont and, 53
schizophrenia, 152, 153
science
  application of, 63
  as medicine, 62
  modernization of, 75

scientists, 5
Sears-Roebuck catalog, 90
seeds, 11
Seeman, Enoch, 60
Seige cycle, 7–8
semisynthetics, 96–98
Senegal, 126
Sertürner, Friedrich, 39–41
sex, 163–86
Shelley, Mary, 77
Shelley, Percy, 27
Sherwill, W. S., 28, 28
  lithograph by, 31
shock, 127
side effects, 7, 128. See also
  effectiveness
  deaths from, 173
  dosage and, 154
  drugs and, 230–31
  Salvarsan and, 102
  statins and, 212
silver, 30
the Sirocco destroyer, 123–26
the sixties, 192, 198
SKF. See Smith, Kline & French
sleep, 13
sleeping pill, 79, 82
the small pox, 50–74
  deaths from, 50
  epidemics of, 50–51, 58
  patients and, 51
  vaccination for, 73
Smith, Kline & French (SKF),
  146
Smith, Nayland, 95
smoking, 30
smugglers, 32
society, 5, 7
  response to, 44
  after World War II, 149
soldiers, 104
Somalia, 70
spices, 14
standardization, 39
  drugs and, 24
starvation, 210
State Research Center of Virology and
  Biotechnology, 72
statins, 211–39
steroids, 170
stimuli, 202

Stoltz, Jamison, 8
strait jackets, 82, 132, *134*
    CPZ and, 150, 155
streptococci, 100
    mice and, 108
    wound infection and, 105
stress, 126. *See also* post-traumatic stress
        disorder
studies, 206
    brain, 156
    for drugs, 23
suicide, 46
    murder-suicide, 47
sulfanilamide (sulfa), 108–22
    azo dyes and, 109
    Domagk and, 117, 255
survivors, 109
Sydenham, Thomas, 24,
        24–25
synthetic dyes, 87
    coal tar from, 103
syphilis, 102
syringe, 42, 177

the Taiping Rebellion, 34
tardive dyskinesia, 154
tea, 30
techniques, 57
technology, 5
temperature, 125
terminology, 3, 9
    addiction and, 91–92
    opium and, 14–15
    vaccination as, 67
terrorism, 73
testing, 106, 109
    animals and, 79, 87
texts, 17, 91
    ancient Islamic and, 20
theories, 69
therapeutic window, 46
Thorazine, 147, 147–50
Thornton, Mary ("Gold Tooth"), 81
Time magazine, 147
tinctures, 24
tobacco, 29
    addiction and, 30
tools, 141
    physicians and, 17, 99
    technology and, 5
TR. *See* Roosevelt, Theodore

trade, 29
    restrictions on, 31
trade names, 97
    antipsychotics and, 157
tranquilizers, 82
travel, 53–54
treatments, 45
    humours and, 58
    mental illness and, 132–46
    physicians and, 51
Trump, Donald, 205
trust, 57
tuberculosis, 26, 106
    patients and, 89
Turkey, 57
Twilight Sleep, 45

United States, 44
    addiction in, 203–5
    Comstock Laws in,
        168–69
    drug subculture in, 35
    life expectancy in, 120
    morphine in, 85
    narcotics control in, 94
    opium use in, 38
    prescriptions and, 1–2
    smallpox in, 68, 73
urea, 77
uses, of drugs, 18, 25
    legitimacy and, 82
    opiates and per capita, 43
    the sixties and, 192, 198
    in United States, 38

vaccines, 4
    antibiotics and, 5
    military and, 73
    renewal of, 71
    for the small pox, 73
    story of, 67
    terminology and, 67
variations
    azo dye, 108
    laudanum and, 37
variola virus, 69
veterans, 198
    addiction and, 198
    PTSD and, 157
    wounds and, 43
Viagra, 181–86

victim blame, 80
viruses, 69
    antibiotics and, 121
    reconstruction of, 73
vitalism, 76
Voltaire, 59
volunteers, 88

Wanrong, 36
wars, 33
    American Civil War, 43
    The Opium Wars, 32–33
    World War I, 97, 104
    World War II, 149, 167
weakness, 231–33
wealth, 153, 168
the White House, 100
    celebrities in, 199
willow bark, 88
Wöhler, Friedrich, 77
women
    education and, 49
    fashion and, 52

morphine and, 44–45
Muslim, 54–55
as nobility, 55
options for, 1
prisoners, 153
rights for, 167
testing and, 109
the World Health Organization, 72
World War I, 97
    Domagk during, 104
World War II, 167
    society after, 149
    soldiers in, 104
wounds, 104
    infections and, 105,
        117
    veterans and, 43
Wren, Christopher, 25
writing, 64
    reputation and, 54

Zauberkugeln (magic balls),
    101